SAS/STAT® 9.22 User's Guide
Statistical Graphics Using ODS
(Book Excerpt)

SAS® Documentation

The correct bibliographic citation for this manual is as follows: SAS Institute Inc. 2010. *SAS/STAT® 9.22 User's Guide: Statistical Graphics Using ODS (Book Excerpt)*. Cary, NC: SAS Institute Inc.

SAS/STAT® 9.22 User's Guide: Statistical Graphics Using ODS (Book Excerpt)

Copyright © 2010, SAS Institute Inc., Cary, NC, USA

ISBN 978-1-60764-632-7

All rights reserved. Produced in the United States of America. No part of this publication may be reproduced, stored in a retrieval system, or transmitted, in any form or by any means, electronic, mechanical, photocopying, or otherwise, without the prior written permission of the publisher, SAS Institute Inc.

U.S. Government Restricted Rights Notice: Use, duplication, or disclosure of this software and related documentation by the U.S. government is subject to the Agreement with SAS Institute and the restrictions set forth in FAR 52.227-19, Commercial Computer Software-Restricted Rights (June 1987).

SAS Institute Inc., SAS Campus Drive, Cary, North Carolina 27513.

1st electronic book, May 2010
1st printing, June 2010

SAS® Publishing provides a complete selection of books and electronic products to help customers use SAS software to its fullest potential. For more information about our e-books, e-learning products, CDs, and hard-copy books, visit the SAS Publishing Web site at **support.sas.com/publishing** or call 1-800-727-3228.

SAS® and all other SAS Institute Inc. product or service names are registered trademarks or trademarks of SAS Institute Inc. in the USA and other countries. ® indicates USA registration.

Other brand and product names are trademarks of their respective companies.

Chapter 21
Statistical Graphics Using ODS

Contents

Introduction	**607**
Getting Started with ODS Statistical Graphics	**609**
Default Plots for Simple Linear Regression with PROC REG	609
Survival Estimate Plot with PROC LIFETEST	612
Contour and Surface Plots with PROC KDE	613
Contour Plots with PROC KRIGE2D	615
Partial Least Squares Plots with PROC PLS	618
Box-Cox Transformation Plot with PROC TRANSREG	620
LS-Means Diffogram with PROC GLIMMIX	621
Principal Component Analysis Plots with PROC PRINCOMP	623
Grouped Scatter Plot with PROC SGPLOT	625
A Primer on ODS Statistical Graphics	**627**
Graph Styles	628
ODS Destinations	631
Accessing Individual Graphs	632
Specifying the Size and Resolution of Graphs	633
Modifying Your Graphs	634
Procedures That Support ODS Graphics	637
Procedures That Support ODS Graphics and Traditional Graphics	637
Syntax	**638**
ODS GRAPHICS Statement	638
ODS Destination Statements	641
PLOTS= Option	642
Selecting and Viewing Graphs	**644**
Specifying an ODS Destination for Graphics	644
Viewing Your Graphs in the SAS Windowing Environment	646
Determining Graph Names and Labels	647
Selecting and Excluding Graphs	649
Graphics Image Files	**651**
Image File Types	651
Naming Graphics Image Files	652
Saving Graphics Image Files	654
Creating Graphs in Multiple Destinations	655
Graph Size and Resolution	**656**

ODS Graphics Editor	**658**
Enabling the Creation of Editable Graphs	658
Editing a Graph with the ODS Graphics Editor	659
The Default Template Stores and the Template Search Path	**662**
Styles	**664**
An Overview of Styles	664
Style Elements and Attributes	666
Style Definitions and Colors	667
Some Common Style Elements	668
Creating an All-Color Style by Using the ModStyle Macro	675
Changing the Default Markers and Lines	677
Changing the Default Style	685
Graph Templates	**687**
The Graph Template Language	687
Locating Templates	692
Displaying Templates	693
Editing Templates	695
Saving Customized Templates	697
Using Customized Templates	697
Reverting to the Default Templates	698
Template Modification Macros	**699**
Style Template Modification Macro	699
Graph Template Modification Macro	701
Adding a BY Line to Graphs	706
Statistical Graphics Procedures	**708**
The SGPLOT Procedure	708
The SGSCATTER Procedure	709
The SGPANEL Procedure	711
The SGRENDER Procedure	714
Examples of ODS Statistical Graphics	**719**
Example 21.1: Creating Graphs with Tool Tips in HTML	719
Example 21.2: Creating Graphs for a Presentation	720
Example 21.3: Creating Graphs in PostScript Files	722
Example 21.4: Displaying Graphs Using the DOCUMENT Procedure	724
Example 21.5: Customizing Graphs Through Template Changes	728
Modifying Graph Titles and Axis Labels	728
Modifying Colors, Line Styles, and Markers	733
Modifying Tick Marks and Grid Lines	736
Modifying the Style to Show Grid Lines	737
Example 21.6: Customizing Survival Plots	740
Modifying the Plot Title	741
Modifying the Axes, Legend, and Inset Table	743
Modifying the Layout and Adding a New Inset Table	745
Example 21.7: Customizing Panels	751

Example 21.8: Customizing Axes and Reference Lines	755
Example 21.9: Customizing the Style for Box Plots	762
Example 21.10: Adding Text to Every Graph	765
Adding a Date and Project Stamp to a Few Graphs	767
Adding Data Set Information to a Graph	770
Adding a Date and Project Stamp to All Graphs	771
Example 21.11: PROC TEMPLATE Statement Order and Primary Plots	772
References	**777**

Introduction

Effective graphics are indispensable for modern statistical analysis. They reveal patterns, differences, and uncertainty that are not readily apparent in tabular output. Graphics provoke questions that stimulate deeper investigation, and they add visual clarity and rich content to reports and presentations.

In earlier SAS releases, creating graphs with statistical procedures typically required additional programming steps such as creating output data sets with the values to plot, modifying these data sets with a DATA step program, and using traditional SAS/GRAPH procedures to produce the plots.

SAS 9.2 eliminates the need for additional programming by providing new functionality, referred to as ODS Statistical Graphics (or ODS Graphics for short). ODS Graphics is an extension of ODS (the Output Delivery System). ODS manages procedure output and lets you display it in a variety of destinations, such as HTML and RTF. With ODS Graphics, statistical procedures now produce graphs as automatically as they produce tables, and graphs are now integrated with tables in the ODS output. ODS Graphics is available in procedures in SAS/STAT, Base SAS, SAS/ETS, SAS/QC, SAS/GRAPH, and other products (see the section "Procedures That Support ODS Graphics" on page 637). Note that SAS/GRAPH software is required for ODS Graphics functionality.

ODS Graphics is enabled when you specify the following statement:

```
ods graphics on;
```

When ODS Graphics is enabled, procedures that support ODS Graphics create appropriate graphs, either by default or when you specify procedure options for requesting specific graphs. These options are documented in the "Syntax" section of each procedure chapter, and the "Details" section of each chapter provides an "ODS Graphics" subsection that lists the graphs that are available. Once ODS Graphics is enabled, it stays enabled for the duration of your SAS session. Alternatively, you can turn it off as follows:

```
ods graphics off;
```

For example, you might consider disabling ODS Graphics if your goal is solely to produce computational results. Often though, you can specify ODS GRAPHICS ON and then leave it on. Throughout this chapter, ODS Graphics is enabled only once per section.

This chapter provides a basic introduction to ODS Graphics along with more detailed information. The following list provides a guide to reading this chapter:

- If you want to see a few of the many graphs that are produced by statistical procedures by using ODS Graphics, see the section "Getting Started with ODS Statistical Graphics" on page 609.

- If you are using ODS Graphics for the first time, read the section "A Primer on ODS Statistical Graphics" on page 627, which provides the minimum information that you need to get started.

- If you need to create plots of raw data or your own customized plots of statistical results, see the section "Statistical Graphics Procedures" on page 708, which describes new SAS/GRAPH procedures that use ODS Graphics.

- If you need information about specialized topics such as accessing your graphs, making changes to your graphs, and working with ODS styles, see the detailed discussions starting with the section "Syntax" on page 638 and including the section "Examples of ODS Statistical Graphics" on page 719.

If you are unfamiliar with ODS, see Chapter 20, "Using the Output Delivery System." For complete documentation about the Output Delivery System, see the *SAS Output Delivery System: User's Guide*. For complete documentation about ODS graph templates, see the *SAS/GRAPH: Graph Template Language User's Guide* and the *SAS/GRAPH: Graph Template Language Reference*. For complete documentation about the Graphics Editor, see the *SAS/GRAPH: ODS Graphics Editor User's Guide*. Also see the *SAS/GRAPH: Statistical Graphics Procedures Guide* for information about the statistical graphics procedures.

Getting Started with ODS Statistical Graphics

This section provides examples that illustrate the most basic uses of ODS Graphics with a few of the many plots that are produced by statistical procedures. These examples also illustrate several ODS styles that are useful for statistical analysis.

Default Plots for Simple Linear Regression with PROC REG

This example is taken from the section "Getting Started: REG Procedure" on page 6150 of Chapter 74, "The REG Procedure." It uses the following data from a study of 19 children:

```
data Class;
   input Name $ Height Weight Age @@;
   datalines;
Alfred   69.0 112.5 14   Alice    56.5  84.0 13   Barbara 65.3  98.0 13

... more lines ...

;
```

A larger version of this data set is available in the SASHELP library, and later examples use this data set by specifying `data=sashelp.class`.

The following statements use PROC REG to fit a simple linear regression model in which Weight is the response variable and Height is the independent variable:

```
ods graphics on;

proc reg data=Class;
   model Weight = Height;
run; quit;
```

The ODS GRAPHICS ON statement is specified to request ODS Graphics in addition to the usual tabular output. The statement ODS GRAPHICS OFF is not used here, but it can be specified to disable ODS Graphics.

The graphical output consists of a fit diagnostics panel, a residual plot, and a fit plot. These plots are integrated with the tabular output and are shown in Figure 21.1, Figure 21.2, and Figure 21.3, respectively. The results are displayed in the STATISTICAL style.

Figure 21.1 Fit Diagnostics Panel

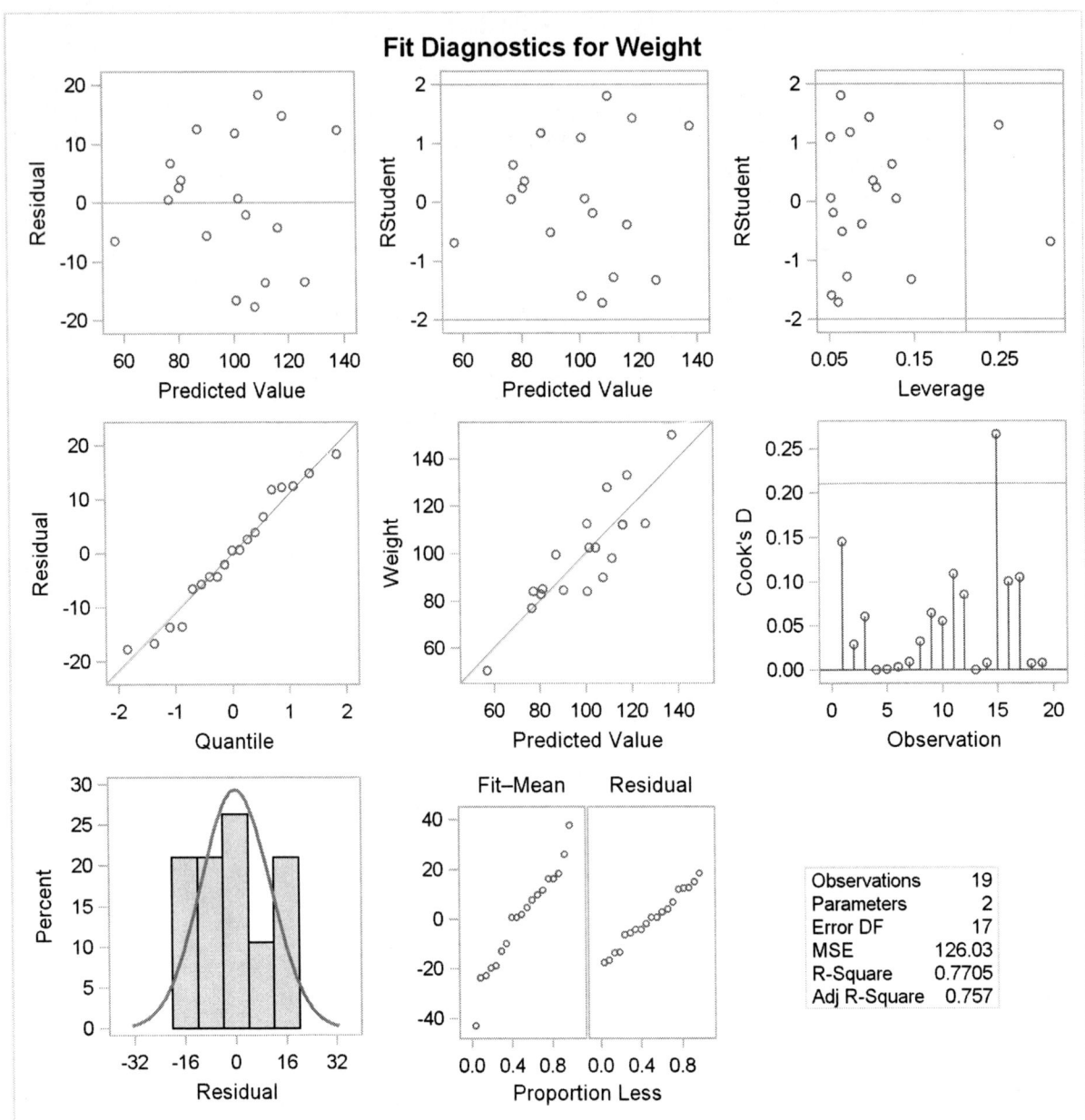

Figure 21.2 Residual Plot

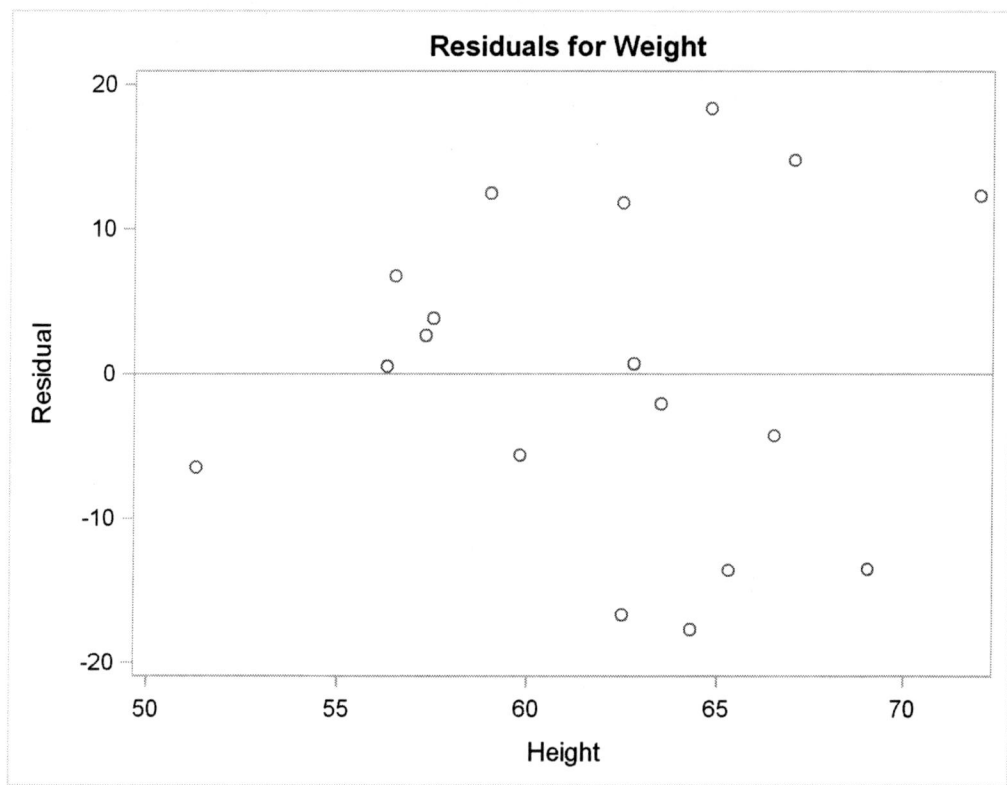

Figure 21.3 Fit Plot

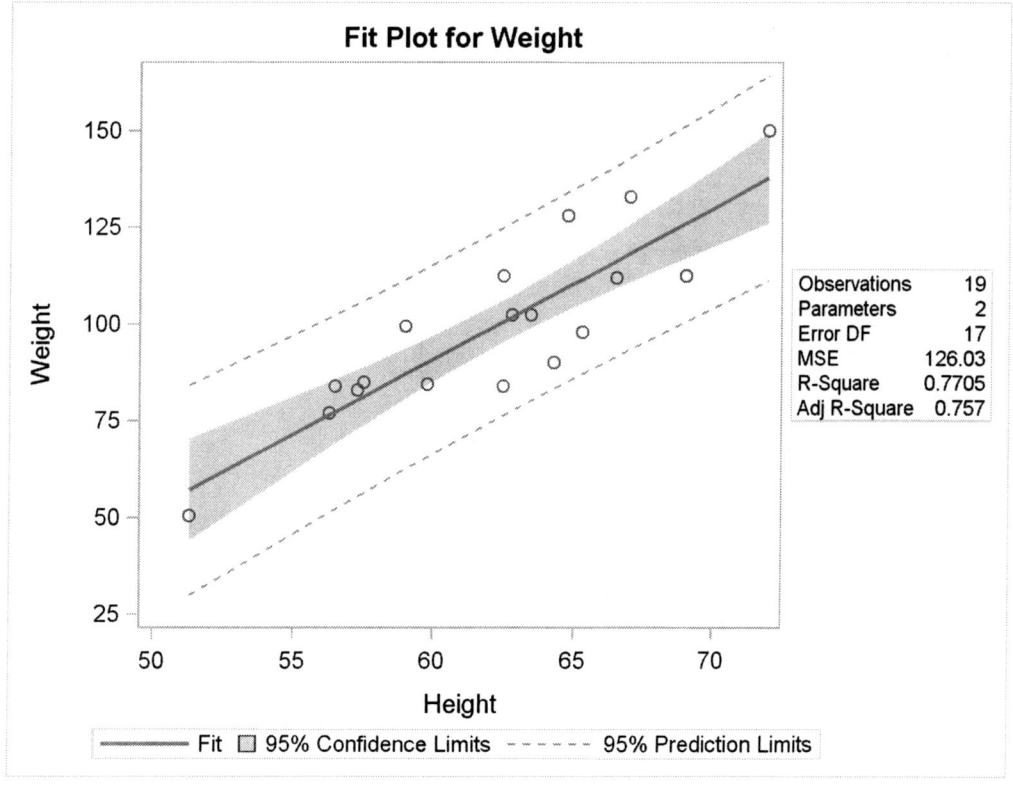

ODS styles control the colors and general appearance of all graphs and tables, and the SAS system provides several styles that are recommended for use with statistical graphics. The STATISTICAL style is the default style in SAS/STAT documentation and is used unless another style is explicitly specified. The default style that you see when you run SAS depends on the ODS destination. Specifically, the default style for the LISTING destination is LISTING, the default style for the HTML destination is DEFAULT, and the default style for the RTF destination is RTF. These and other styles are shown in this chapter. For more information about styles, see the section "Graph Styles" on page 628 and the section "Styles" on page 664.

Survival Estimate Plot with PROC LIFETEST

This example is taken from Example 49.2 of Chapter 49, "The LIFETEST Procedure." It shows how to construct a product-limit survival estimate plot. Both the ODS GRAPHICS statement and procedure options are used to request the plot.

The following statements create a SAS data set with disease-free times for three risk categories:

```
proc format;
   value risk 1='ALL' 2='AML-Low Risk' 3='AML-High Risk';
run;

data BMT;
   input Group T Status @@;
   format Group risk.;
   label T='Disease Free Time';
   datalines;
1 2081 0 1 1602 0 1 1496 0 1 1462 0 1 1433 0

   ... more lines ...

;
```

The following statements use PROC LIFETEST to compute the product-limit estimate of the survivor function for each risk category:

```
ods graphics on;

proc lifetest data=BMT plots=survival(cb=hw test);
   time T * Status(0);
   strata Group / test=logrank;
run;
```

The ODS GRAPHICS ON statement requests ODS Graphics functionality, and the PLOTS=SURVIVAL option requests a plot of the estimated survival curves. The CB=HW suboption requests Hall-Wellner confidence bands, and the TEST suboption displays the p-value for the log-rank test in a plot inset.

Figure 21.4 displays the plot; note that tabular output is not shown. The results are displayed in the STATISTICAL style. Patients in the AML-Low Risk group are disease-free longer than those in the

ALL group, who in turn fare better than those in the AML-High Risk group.

Figure 21.4 Survival Plot

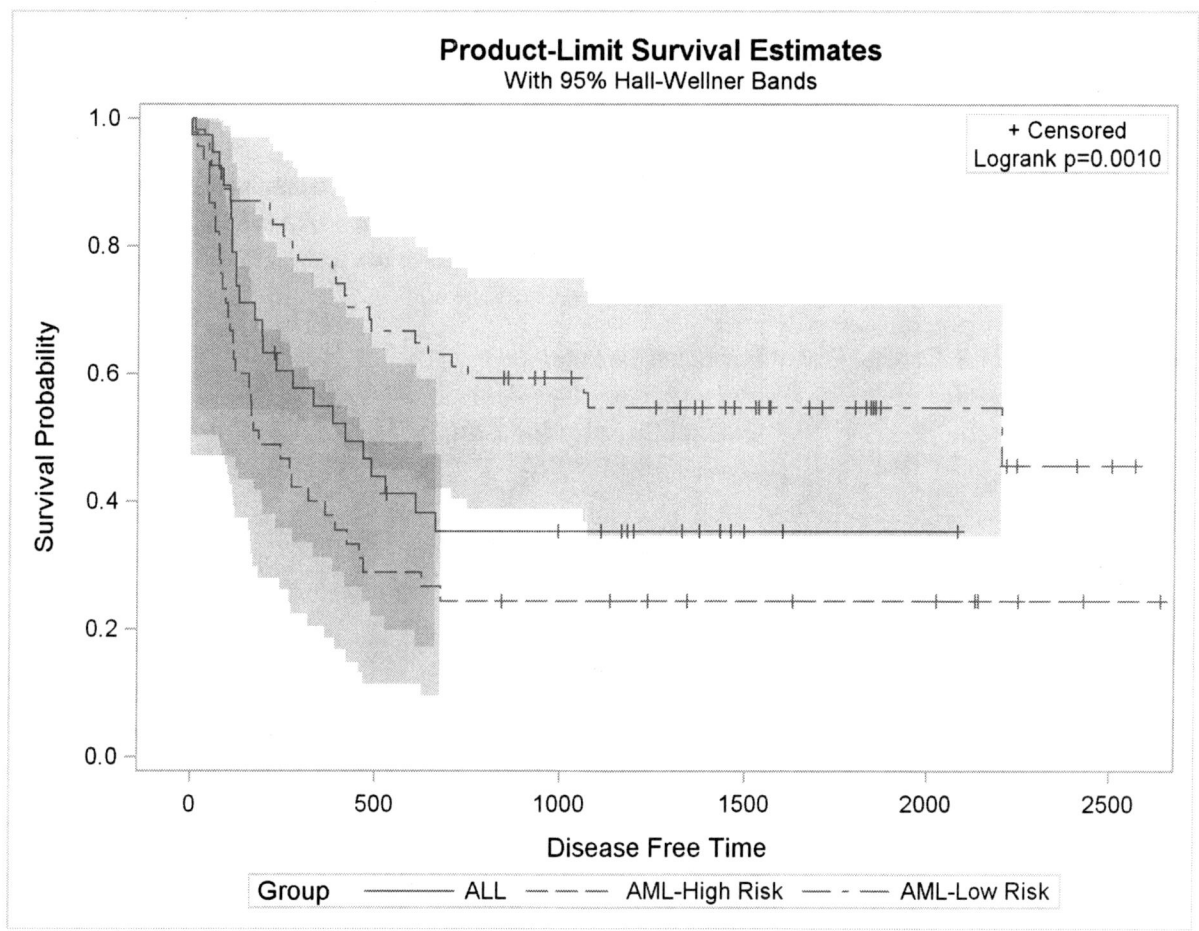

Contour and Surface Plots with PROC KDE

This example is taken from the section "Getting Started: KDE Procedure" on page 3460 in Chapter 45, "The KDE Procedure." Here, in addition to the ODS GRAPHICS statement, procedure options are used to request plots. The following statements simulate 1,000 observations from a bivariate normal density with means (0,0), variances (10,10), and covariance 9:

```
data bivnormal;
   do i = 1 to 1000;
      z1 = rannor(104);
      z2 = rannor(104);
      z3 = rannor(104);
      x  = 3*z1+z2;
      y  = 3*z1+z3;
      output;
   end;
run;
```

The following statements request a bivariate kernel density estimate for the variables x and y:

```
ods graphics on;

proc kde data=bivnormal;
   bivar x y / plots=contour surface;
run;
```

The PLOTS= option requests a contour plot and a surface plot of the estimate (displayed in Figure 21.5 and Figure 21.6, respectively). The results are displayed in the STATISTICAL style. For more information about the graphics available in PROC KDE, see the section "ODS Graphics" on page 3478 of Chapter 45, "The KDE Procedure."

Figure 21.5 Contour Plot of Estimated Density

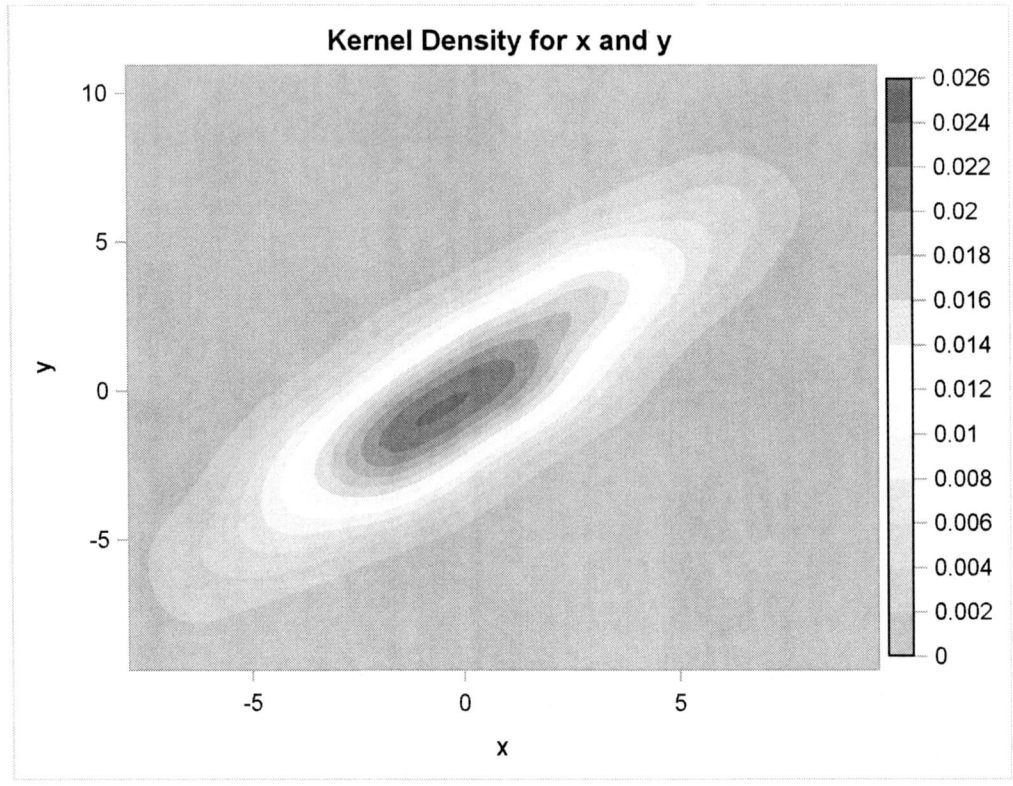

Figure 21.6 Surface Plot of Estimated Density

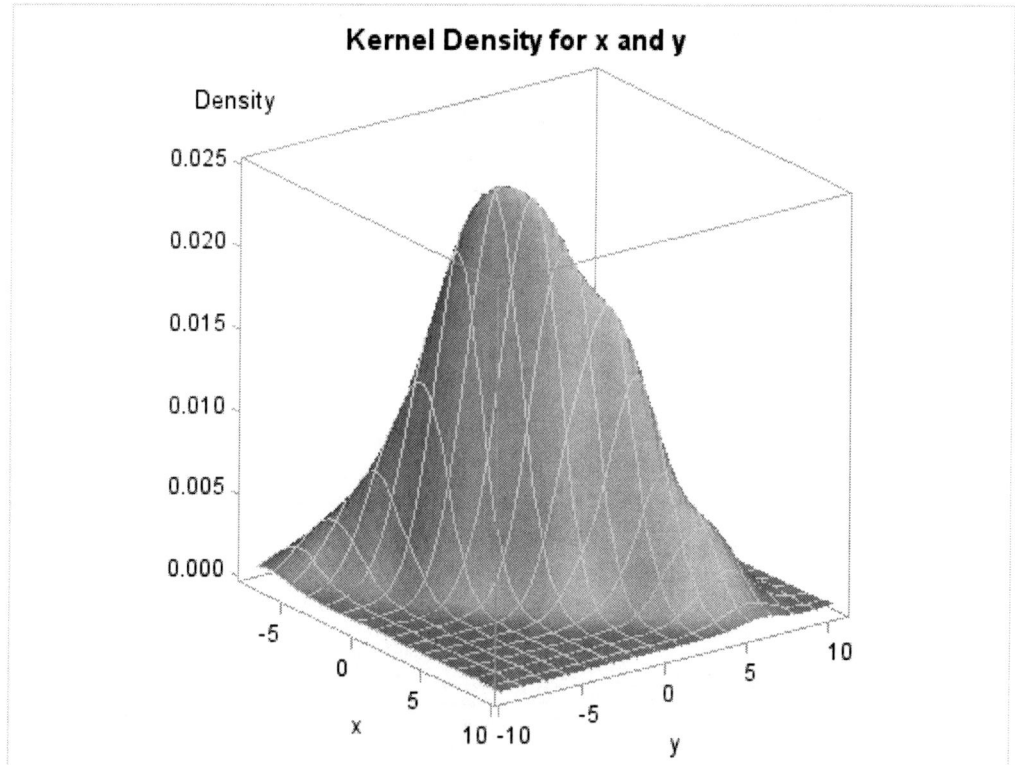

Contour Plots with PROC KRIGE2D

This example is taken from Example 46.2 of Chapter 46, "The KRIGE2D Procedure." The following statements create a SAS data set that contains measurements of coal seam thickness:

```
data thick;
   input East North Thick @@;
   label Thick='Coal Seam Thickness';
   datalines;
 0.7   59.6   34.1     2.1   82.7   42.2     4.7   75.1   39.5

   ... more lines ...

;
```

The following statements set the output style to DEFAULT and run PROC KRIGE2D:

```
ods listing style=default;
ods graphics on;

proc krige2d data=thick outest=predictions
             plots=(observ(showmissing)
                    pred(fill=pred line=pred obs=linegrad)
                    pred(fill=se line=se obs=linegrad));
   coordinates xc=East yc=North;
   predict var=Thick r=60;
   model scale=7.2881 range=30.6239 form=gauss;
   grid x=0 to 100 by 2.5 y=0 to 100 by 2.5;
run;
```

The PLOTS=OBSERV(SHOWMISSING) option produces a scatter plot of the data along with the locations of any missing data. The PLOTS=PRED option produces maps of the kriging predictions and standard errors. Two instances of the PLOTS=PRED option are specified with suboptions that customize the plots. The results are shown in Figure 21.7.

Figure 21.7 PROC KRIGE2D Results Using the DEFAULT Style

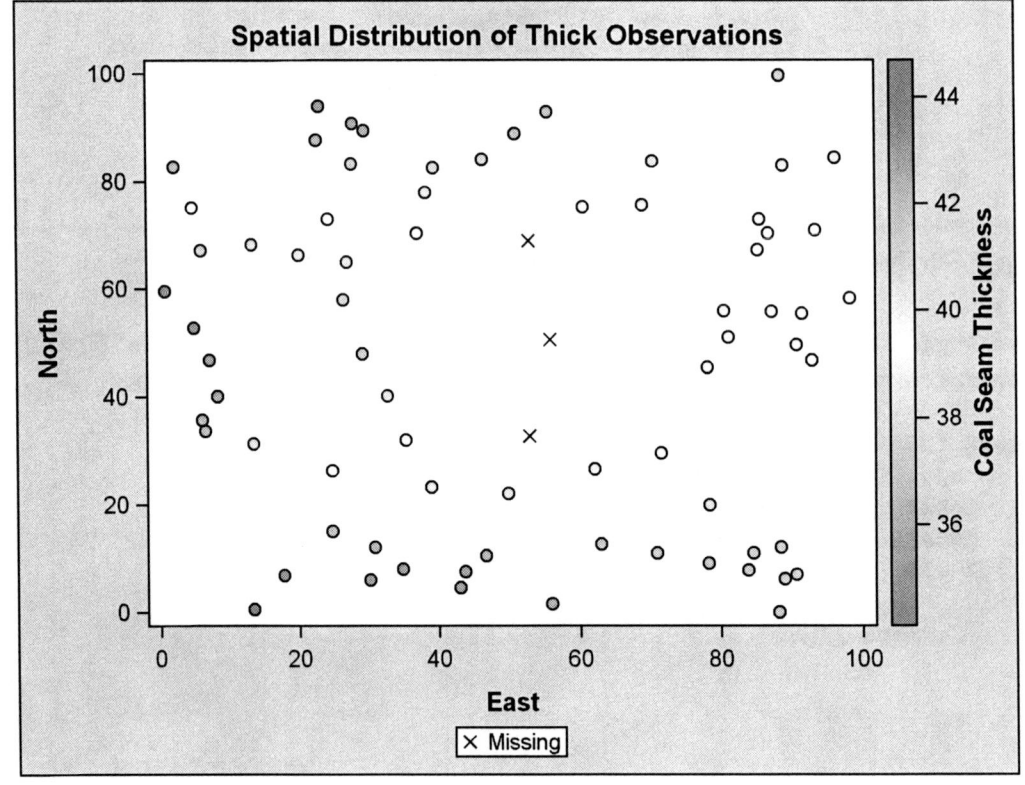

Figure 21.7 continued

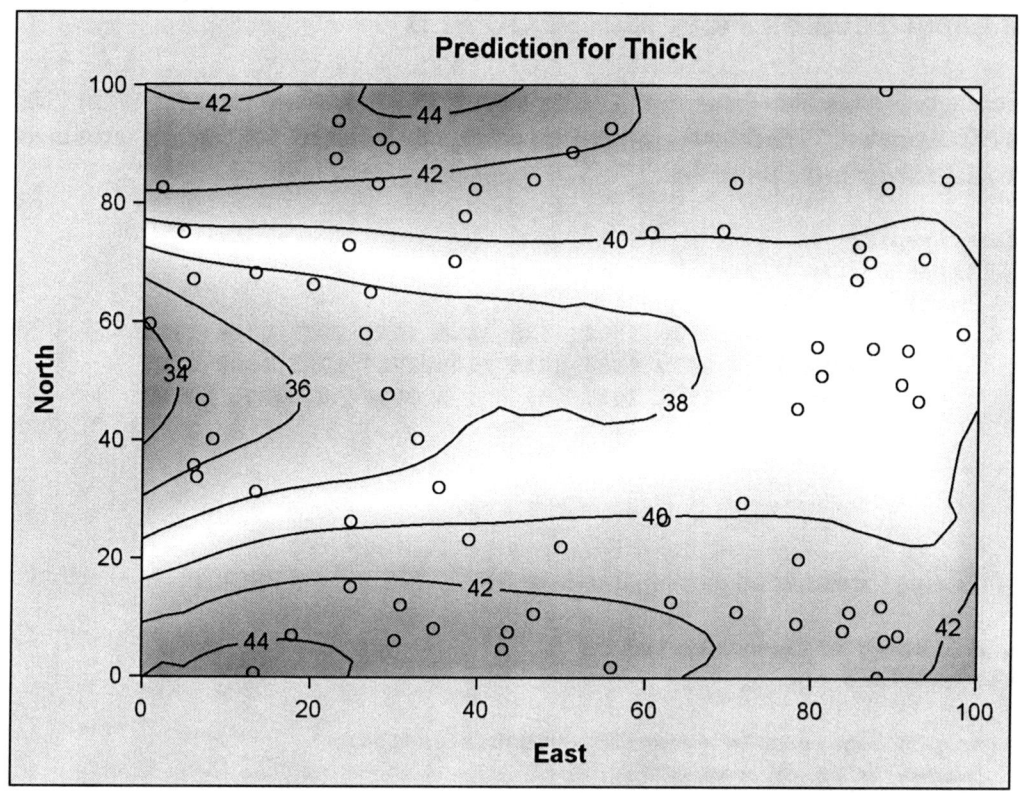

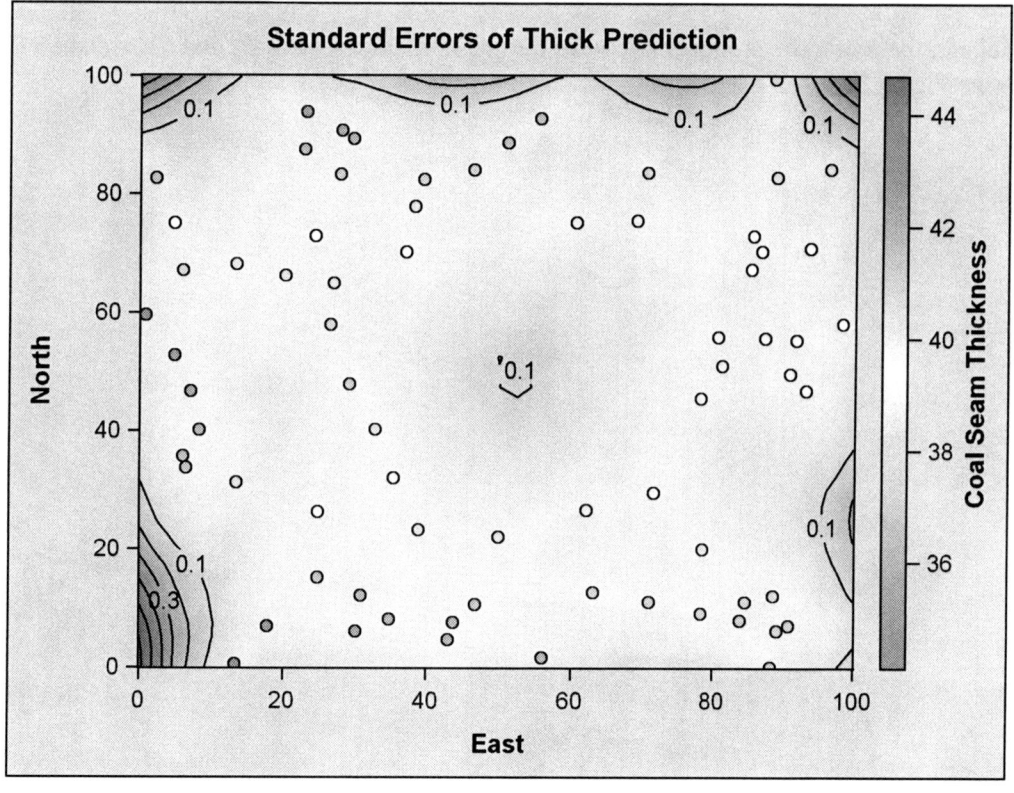

Partial Least Squares Plots with PROC PLS

This example is taken from the section "Getting Started: PLS Procedure" on page 5467 of Chapter 67, "The PLS Procedure." The following statements create a SAS data set that contains measurements of biological activity in the Baltic Sea:

```
data Sample;
   input obsnam $ v1-v27 ls ha dt @@;
   datalines;
EM1    2766 2610 3306 3630 3600 3438 3213 3051 2907 2844 2796
       2787 2760 2754 2670 2520 2310 2100 1917 1755 1602 1467
       1353 1260 1167 1101 1017          3.0110  0.0000   0.00

   ... more lines ...

;
```

The following statements set the output style to ANALYSIS and run PROC PLS:

```
ods listing style=analysis;
ods graphics on;

proc pls data=sample cv=split cvtest(seed=104);
   model ls ha dt = v1-v27;
run;
```

By default, the procedure produces a plot for the cross validation analysis and a correlation loading plot (see Figure 21.8).

Figure 21.8 Partial Least Squares Results Using the ANALYSIS Style

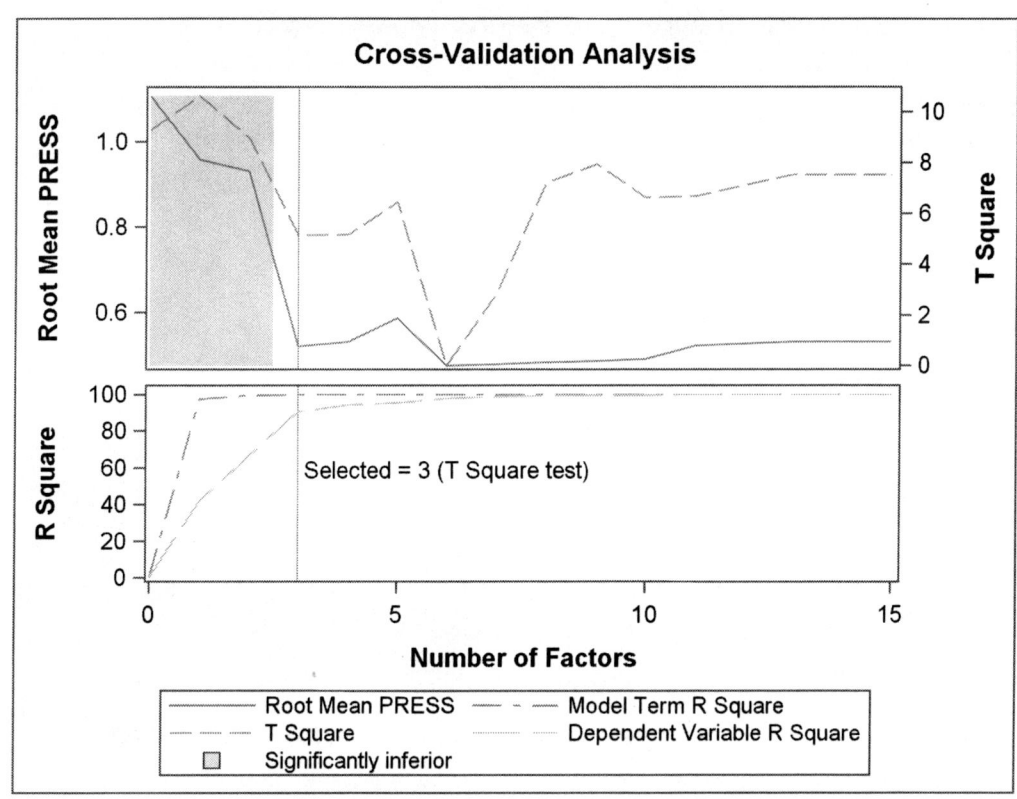

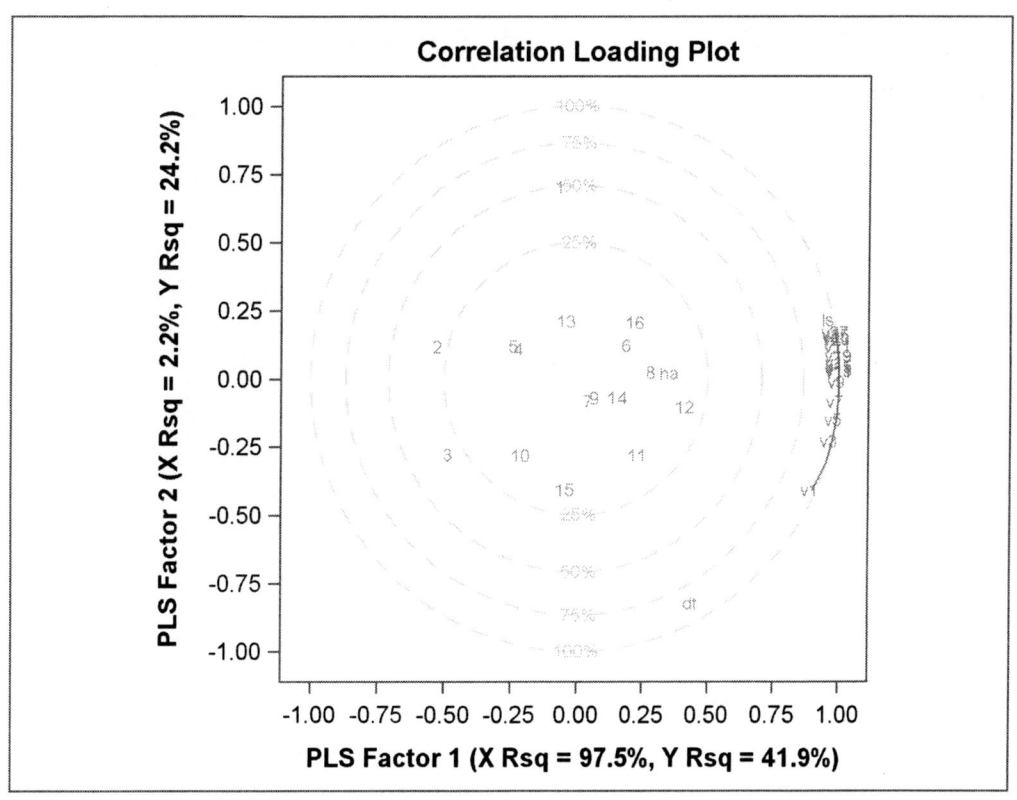

Box-Cox Transformation Plot with PROC TRANSREG

This example is taken from Example 91.2 of Chapter 91, "The TRANSREG Procedure." The following statements create a SAS data set that contains failure times for yarn:

```
proc format;
   value a -1 =   8  0 =   9  1 =  10;
   value l -1 = 250  0 = 300  1 = 350;
   value o -1 =  40  0 =  45  1 =  50;
run;

data yarn;
   input Fail Amplitude Length Load @@;
   format amplitude a. length l. load o.;
   label fail = 'Time in Cycles until Failure';
   datalines;
 674 -1 -1 -1    370 -1 -1  0    292 -1 -1  1    338  0 -1 -1

   ... more lines ...

;
```

The following statements set the output style to JOURNAL2 and run PROC TRANSREG:

```
ods listing style=journal2;
ods graphics on;

proc transreg data=yarn;
   model BoxCox(fail / convenient lambda=-2 to 2 by 0.05) =
         qpoint(length amplitude load);
run;
```

The log-likelihood plot in Figure 21.9 suggests a Box-Cox transformation with $\lambda = 0$.

Figure 21.9 Box-Cox "Significant Effects" Using the JOURNAL2 Style

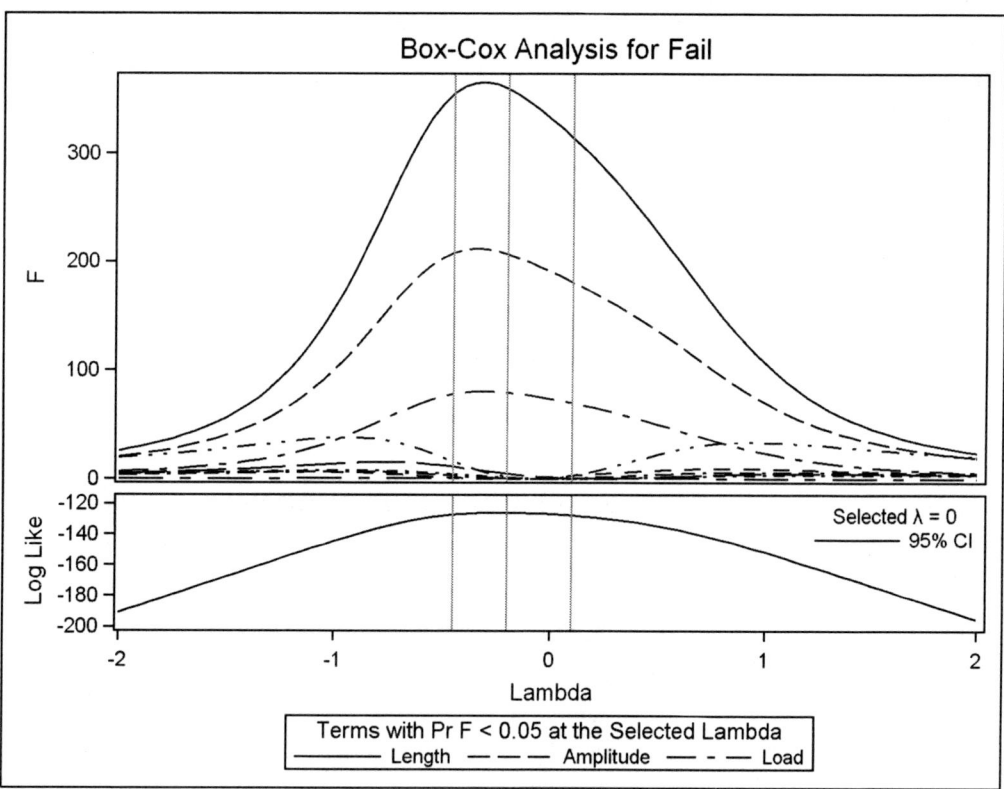

LS-Means Diffogram with PROC GLIMMIX

This example is taken from the section "Graphics for LS-Mean Comparisons" on page 2843 of Chapter 38, "The GLIMMIX Procedure." The following statements create a SAS data set that contains measurements from an experiment that investigates how snapdragons grow in various soils:

```
data plants;
   input Type $ @;
   do Block = 1 to 3;
      input StemLength @;
      output;
   end;
   datalines;
Clarion    32.7 32.3 31.5

   ... more lines ...

;
```

622 ✦ Chapter 21: Statistical Graphics Using ODS

The following statements set the output style to STATISTICAL and run PROC GLIMMIX:

```
ods listing style=statistical;
ods graphics on;

proc glimmix data=plants order=data plots=diffogram;
   class Block Type;
   model StemLength = Block Type;
   lsmeans Type;
run;
```

The PLOTS=DIFFOGRAM option produces a diffogram, shown in Figure 21.10, that displays all of the pairwise least squares mean differences and indicates which are significant.

Figure 21.10 LS-Means Diffogram Using the STATISTICAL Style

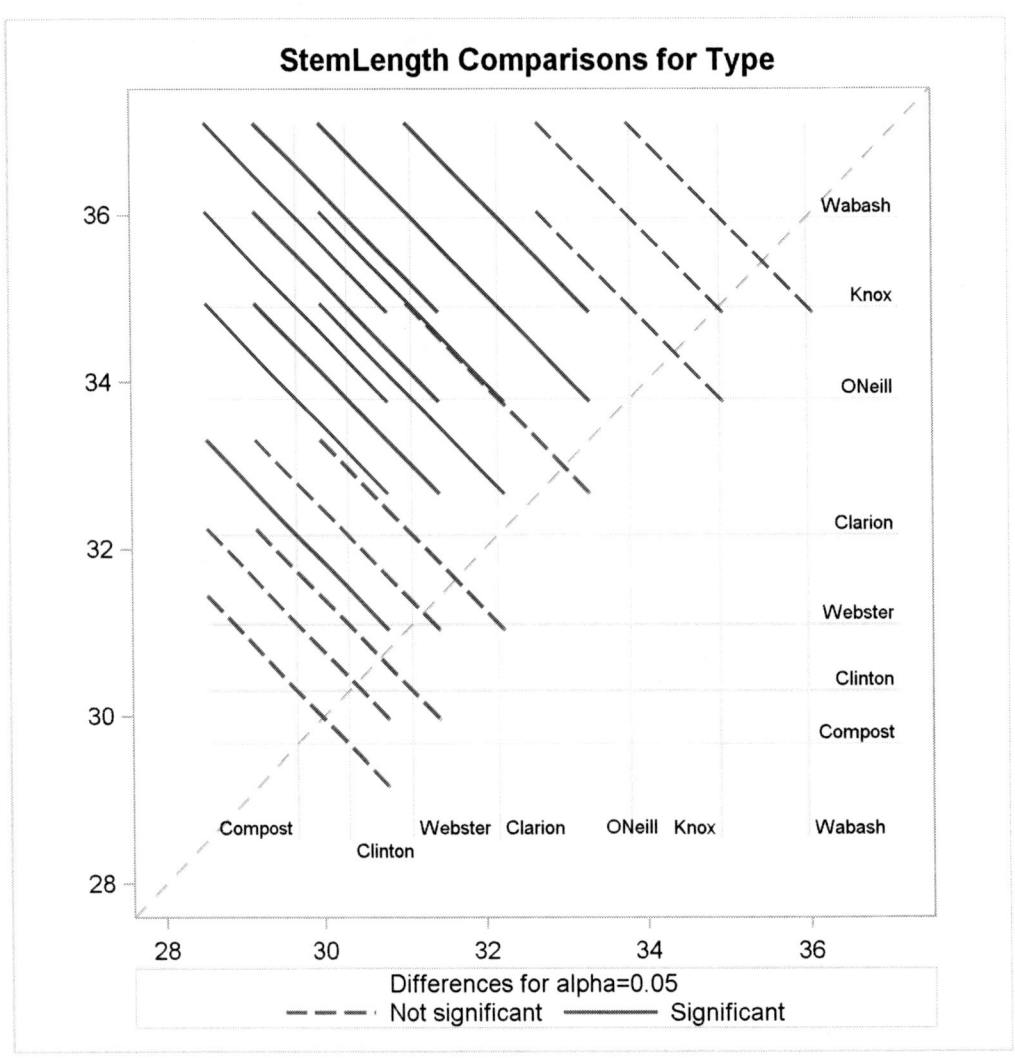

Principal Component Analysis Plots with PROC PRINCOMP

This example is taken from Example 70.3 of Chapter 70, "The PRINCOMP Procedure." The following statements create a SAS data set that contains ratings of job performance of police officers:

```
options validvarname=any;

data Jobratings;
   input ('Communication Skills'n
          'Problem Solving'n
          'Learning Ability'n
          'Judgment Under Pressure'n
          'Observational Skills'n
          'Willingness to Confront Problems'n
          'Interest in People'n
          'Interpersonal Sensitivity'n
          'Desire for Self-Improvement'n
          'Appearance'n
          'Dependability'n
          'Physical Ability'n
          'Integrity'n
          'Overall Rating'n) (1.);
   datalines;
26838853879867

   ... more lines ...

;
```

The following statements set the output style to RTF and run PROC PRINCOMP:

```
ods listing style=rtf;
ods graphics on;

proc princomp data=Jobratings(drop='Overall Rating'n) n=2
              plots=(Matrix PatternProfile);
run;
```

The plots are requested by the PLOTS=(MATRIX PATTERNPROFILE) option. The results, shown in Figure 21.11, contain the default scree and variance-explained plots, along with a scatter plot matrix of component scores and a pattern profile plot.

Figure 21.11 Principal Components Using the RTF Style

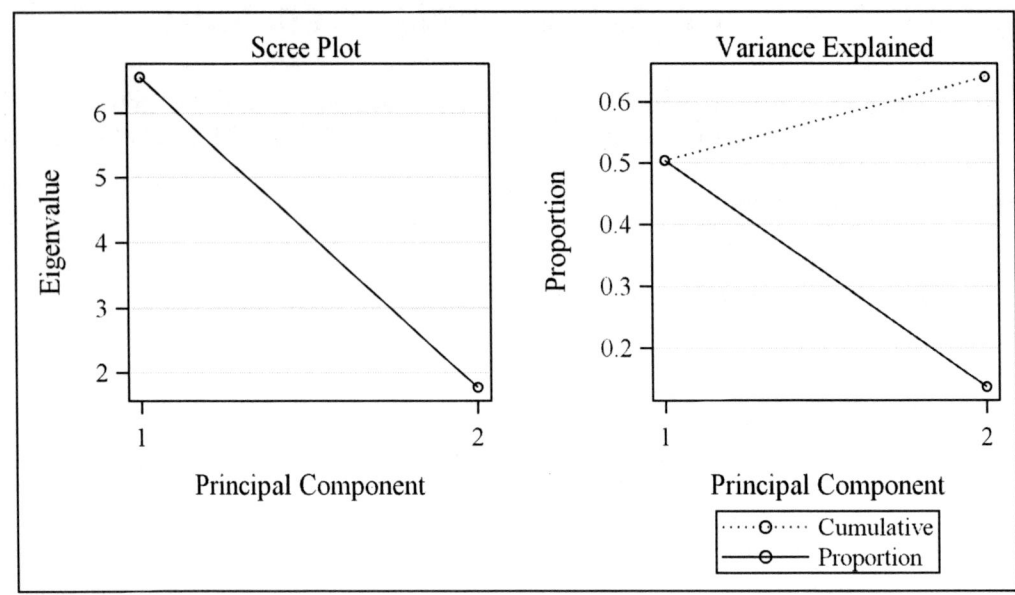
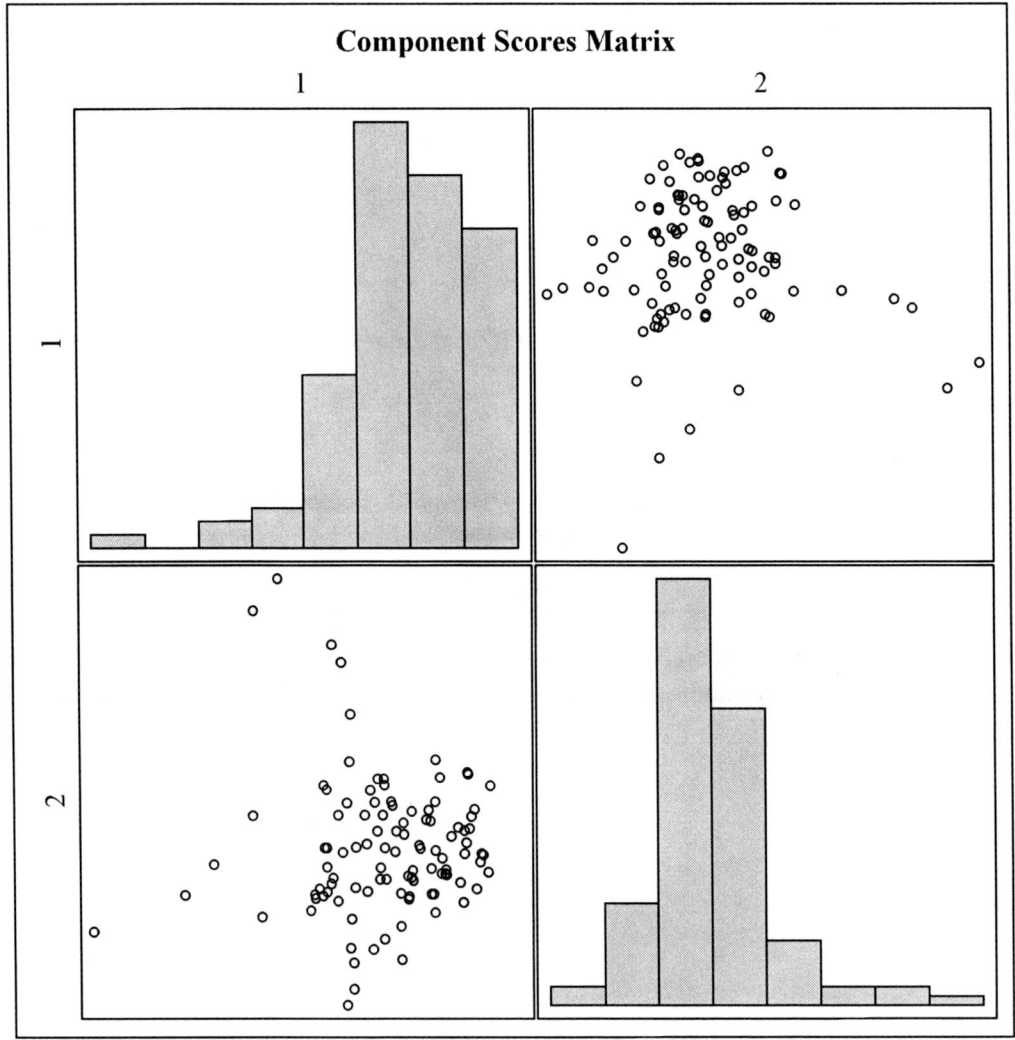

Figure 21.11 *continued*

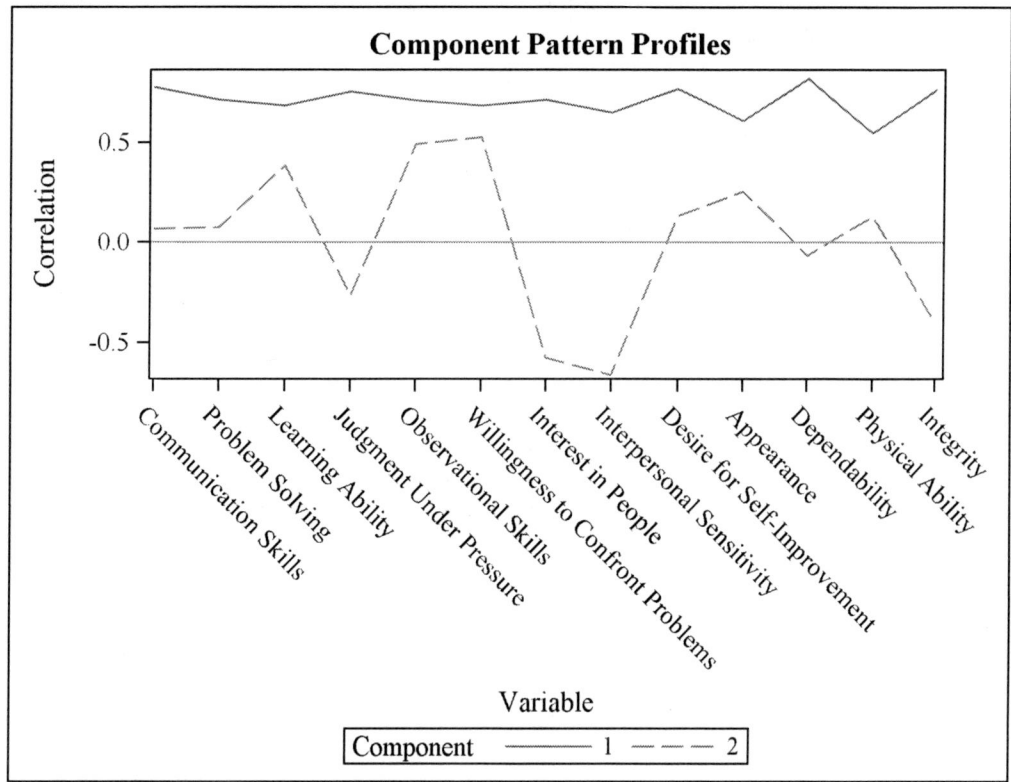

Grouped Scatter Plot with PROC SGPLOT

This example is taken from Example 31.1 of Chapter 31, "The DISCRIM Procedure." The following statements create a SAS data set from the iris data:

```
proc format;
   value specname
      1='Setosa    '
      2='Versicolor'
      3='Virginica ';
run;

data iris;
   input SepalLength SepalWidth PetalLength PetalWidth
         Species @@;
   format Species specname.;
   label SepalLength='Sepal Length in mm.'
         SepalWidth ='Sepal Width in mm.'
         PetalLength='Petal Length in mm.'
         PetalWidth ='Petal Width in mm.';
   datalines;
50 33 14 02 1 64 28 56 22 3 65 28 46 15 2 67 31 56 24 3

   ... more lines ...

;
```

626 ✦ Chapter 21: Statistical Graphics Using ODS

The following statements set the output style to LISTING and run PROC SGPLOT to make a scatter plot, grouped by iris species:

```
ods listing style=listing;

proc sgplot data=iris;
   title 'Fisher (1936) Iris Data';
   scatter x=petallength y=petalwidth / group=species;
run;
```

The results are shown in Figure 21.12.

Figure 21.12 Scatter Plot That Uses the LISTING Style

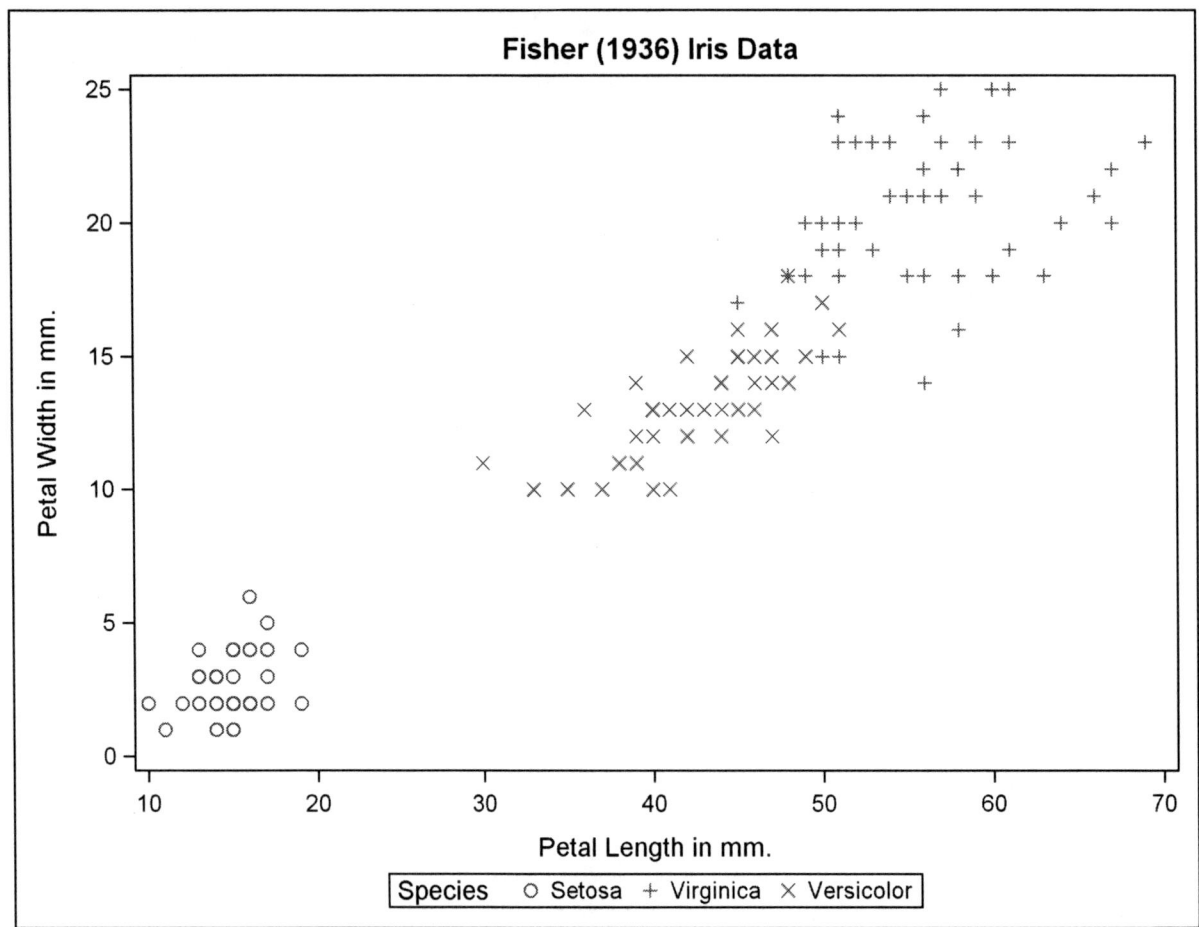

See the section "Statistical Graphics Procedures" on page 708 and the *SAS/GRAPH: Statistical Graphics Procedures Guide* for more information about PROC SGPLOT (statistical graphics plot) and other SG procedures. You do not need to enable ODS Graphics in order to use SG procedures (because making plots with ODS Graphics is their sole function).

A Primer on ODS Statistical Graphics

You invoke ODS Graphics by specifying the following statement:

```
ods graphics on;
```

ODS Graphics remains in effect for all procedure steps until you turn it off with the following statement:

```
ods graphics off;
```

See the section "Syntax" on page 638 for details about the more commonly used ODS GRAPHICS statement options. Once you have invoked ODS Graphics, creating graphical output with procedures is as simple as creating tabular output. You can control your output in the following ways:

- ODS destination statements (such as ODS HTML or ODS RTF) specify where you want your graphs displayed. See Figure 21.13 for an example of HTML output. See the section "ODS Destination Statements" on page 641 for a list of the supported destinations. See the section "Syntax" on page 638 for details about the more commonly used ODS destination statement options.

- ODS SELECT and ODS EXCLUDE statements select and exclude graphs from your output. See the section "Selecting and Excluding Graphs" on page 649 for an example of how to select graphs.

- ODS OUTPUT statements create SAS data sets from the data object used to make the plot. See the section "Specifying an ODS Destination for Graphics" on page 644 for an example.

- Procedure options specify which graphs to create. For each procedure, these options are described in the "Syntax" section of the procedure chapter. Typically, you use the PLOTS= option to control all graphs. The available graphs are listed in the "ODS Graphics" section, which is found in the "Details" section of each procedure chapter. Many graphs are produced by default.

- ODS styles control the general appearance and consistency of all graphs and tables. See the sections "Graph Styles" on page 628 and "Styles" on page 664 for more information about styles.

- ODS templates modify the layout and details of each graph. See the section "Graph Templates" on page 687 for more information about templates.

  NOTE: A default template is provided by SAS for each graph, so you do not need to know anything about templates to create statistical graphics.

You can also access individual graphs, control the resolution and size of graphs, and modify your graphs (as explained in the sections beginning with "Selecting and Viewing Graphs" on page 644). Alternatively, you can use special statistical graphics procedures to create custom graphs directly (see the section "Statistical Graphics Procedures" on page 708).

Graph Styles

ODS styles control the overall appearance of graphs and tables. They specify colors, fonts, line styles, and other attributes of graph elements. The following styles are recommended for statistical work:

- The DEFAULT style is a color style intended for general-purpose work. See Figure 21.13 for an example of the DEFAULT style, which is the default style for the HTML destination. Most other styles inherit some of their elements from this style.

- The STATISTICAL style is a color style recommended for output in Web pages or color print media. The STATISTICAL style might not necessarily print well on black-and-white devices. See Figure 21.14 for an example. This is the default style for SAS/STAT documentation.

- The ANALYSIS style is a color style with a somewhat different appearance from the STATISTICAL style. See Figure 21.8 for an example.

- The JOURNAL family of styles (JOURNAL and JOURNAL2) consists of black-and-white or gray-scale styles that are recommended for graphs that appear in journals and in other black-and-white publications. See Figure 21.15 for an example of the JOURNAL style, see Figure 21.9 for an example of the JOURNAL2 style, and see Example 21.3 for a comparison of the two styles.

- The RTF style is used to produce graphs to insert into a Microsoft Word document or a Microsoft PowerPoint slide. See Figure 21.11 for an example of the RTF style, which is the default style for the RTF destination.

There are many other styles including the LISTING style, which is the default style for the LISTING destination.

Figure 21.13 HTML Output from PROC REG with DEFAULT Style

Figure 21.14 HTML Output from PROC REG with the Statistical Style

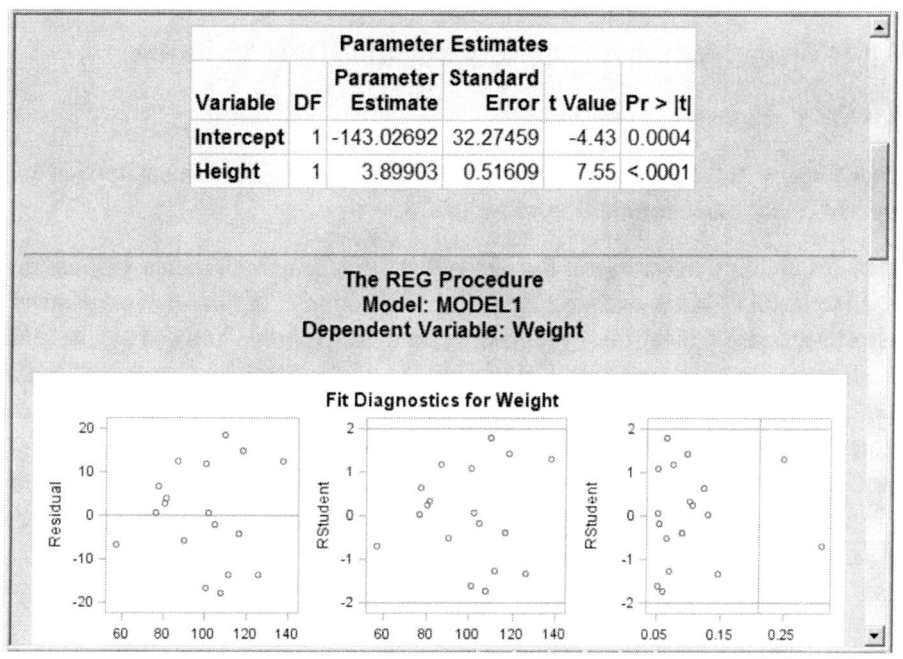

Figure 21.15 HTML Output from PROC REG with the Journal Style

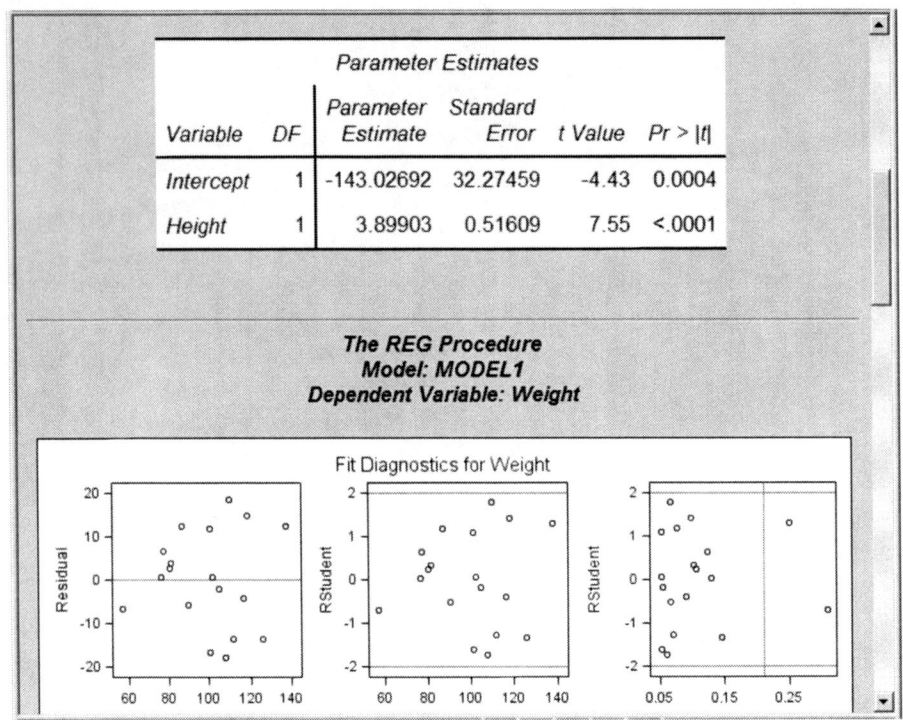

You specify a style with the STYLE= option in the ODS destination statement. For example, the following statement requests HTML output produced with the JOURNAL style:

```
ods html style=Journal;
```

Similarly, the following statement sets the style for the LISTING destination:

```
ods listing style=Statistical;
```

The style specified with the STYLE= option in the ODS LISTING statement applies only to graphs. The legacy SAS monospace format is used for tables.

The color styles are a compromise in the sense that some graph elements are intentionally over-distinguished to facilitate black-and-white printing. For example, fit lines that correspond to different classification levels are distinguished by both colors and line patterns. You can use the %MODSTYLE SAS autocall macro (see the sections "Creating an All-Color Style by Using the ModStyle Macro" on page 675 and "Style Template Modification Macro" on page 699) to modify a style so that it relies only on color for distinguishability. More generally, you can modify the colors, fonts, and other attributes of graph elements in a style by editing the style definition template. More information is provided in the section "Styles" on page 664, and detailed information is in the *SAS Output Delivery System: User's Guide*.

ODS Destinations

ODS can send your graphs and tables to a number of different destinations including RTF (rich text format), HTML (hypertext markup language), LISTING (the SAS LISTING destination), DOCUMENT (the ODS document), and PDF (portable document format). You use an ODS statement to open a destination, as in the following examples:

```
ods html body='b.htm';
ods rtf;
ods listing;
ods document name=MyDoc(write);
ods pdf file="contour.pdf";
```

You can close destinations individually or all at once, as in the following examples:

```
ods html close;
ods rtf close;
ods listing close;
ods document close;
ods pdf close;
ods _all_ close;
```

For most ODS destinations (for example, HTML, RTF, and PDF), graphs and tables are integrated in the output, and you view your output with an appropriate viewer, such as a web browser for HTML. However, the default LISTING destination is different. If you are using the LISTING destination in the SAS windowing environment, you view your graphs individually by clicking the graph icons in the Results window, as illustrated in Figure 21.16. This action invokes a host-dependent graph viewer (for example, Microsoft Photo Editor on Windows). The graphs produced with ODS Graphics are *not* displayed with traditional graphs in the Graph window.

Figure 21.16 SAS Results Window

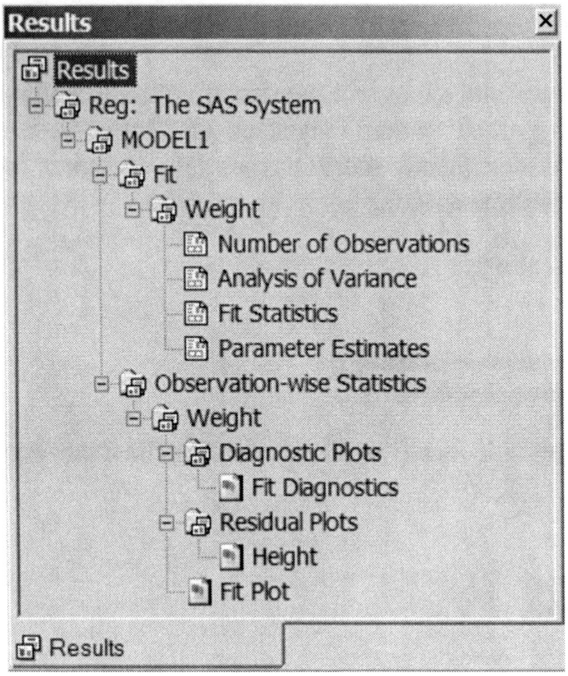

If you are using the SAS windowing environment and you prefer to view integrated output, you should specify a destination such as HTML or RTF. At the same time, you can prevent the Output window from appearing by closing the LISTING destination, as in the following statements:

```
ods listing close;
ods html;
```

A graph is created for every open destination. When you open a new destination, you should close all destinations that you do not need. Closing destinations makes your jobs run faster and with fewer resources, because fewer graphs are produced.

Accessing Individual Graphs

If you are writing a paper or creating a presentation, you need to access your graphs individually. There are various ways to do this, depending on the ODS destination. Three particularly useful methods are as follows:

- If you are viewing RTF output, you can simply copy and paste your graphs from the viewer into a Microsoft Word document or a Microsoft PowerPoint slide.

- If you are viewing HTML output, you can copy and paste your graphs from the viewer, or you can right-click the graph and save it to a file. Copying and pasting from RTF is preferable because the default resolution is higher than with HTML. See the section "Specifying the Size and Resolution of Graphs" on page 633 for details.

- You can save your graphs in image files and then include them into a paper or presentation. For example, you can save your graphs as PNG files and include them into a paper that you are writing with LaTeX or into an HTML document.

You can specify the graphics image format and the file name in the ODS GRAPHICS statement. For example, the following statements, when submitted before a procedure step that produces multiple graphs, save the graphs in PostScript files named *myname.ps*, *myname1.ps*, and so on:

```
ods listing close;
ods latex;
ods graphics on / imagefmt=ps imagename='myname';
```

See the section "Image File Types" on page 651 for details about the file types available with various destinations, how they are named, and how they are saved.

If you are using the LISTING destination and the SAS windowing environment, you can also copy from the default viewer into a Microsoft Word document or a Microsoft PowerPoint slide.

Specifying the Size and Resolution of Graphs

Two factors to consider when you are creating graphs for a paper or presentation are the size of the graph and its resolution. You can specify the size of a graph in the ODS GRAPHICS statement. The following examples show typical ways to change the size of your graphs:

```
ods graphics on / width=6in;
ods graphics on / height=4in;
ods graphics on / width=4.5in height=3.5in;
```

You can change the resolution with the IMAGE_DPI= option in any ODS destination statement, as in the following example:

```
ods html image_dpi=300;
```

The default resolution of graphs created with the HTML and LISTING destinations is 100 DPI (dots per inch), whereas the default with the RTF destination is 200 DPI. An increase in resolution often improves the quality of the graphs, but it also increases the size of the image file. See the section "Graph Size and Resolution" on page 656 for more information about graph size and resolution.

Modifying Your Graphs

Although ODS Graphics is designed to automate the creation of high-quality statistical graphics, on occasion you might need to modify your graphs. There are two ways you can make modifications, depending on whether the changes you want to make are data-dependent and immediate (for a specific graph you are preparing for a paper or presentation), or whether they are persistent (applied to a graph each time you run the procedure). You can make immediate, ad hoc changes by using the ODS Graphics Editor, which provides a point-and-click interface. You can make persistent changes by modifying the ODS graph template for a particular plot. A graph template is a program, written in the Graph Template Language (GTL), that specifies the layout and details of a graph.

NOTE: The SAS system provides a template for each graph it creates, so you do not need to know anything about templates to create statistical graphics.

You can use the ODS Graphics Editor to customize titles and labels, annotate data points, add text, and change the properties of graph elements. After you have modified your graph, you can save it as a PNG image file or as an SGE file, which preserves the editing context. You can open SGE files with the ODS Graphics Editor and resume editing.

You can invoke the ODS Graphics Editor in the SAS windowing environment, provided that the LISTING destination is open and that you have enabled ODS Graphics to create editable graphs. The steps for doing this are described in the section "ODS Graphics Editor" on page 658. Also see *SAS/GRAPH: ODS Graphics Editor User's Guide*.

Figure 21.17 shows the ODS Graphics Editor window for a fit plot created by PROC REG. Figure 21.18 shows modifications made with tools in the ODS Graphics Editor. The title has been changed, and the legend has been repositioned.

Figure 21.17 ODS Graphics Editor Invoked with a Fit Plot

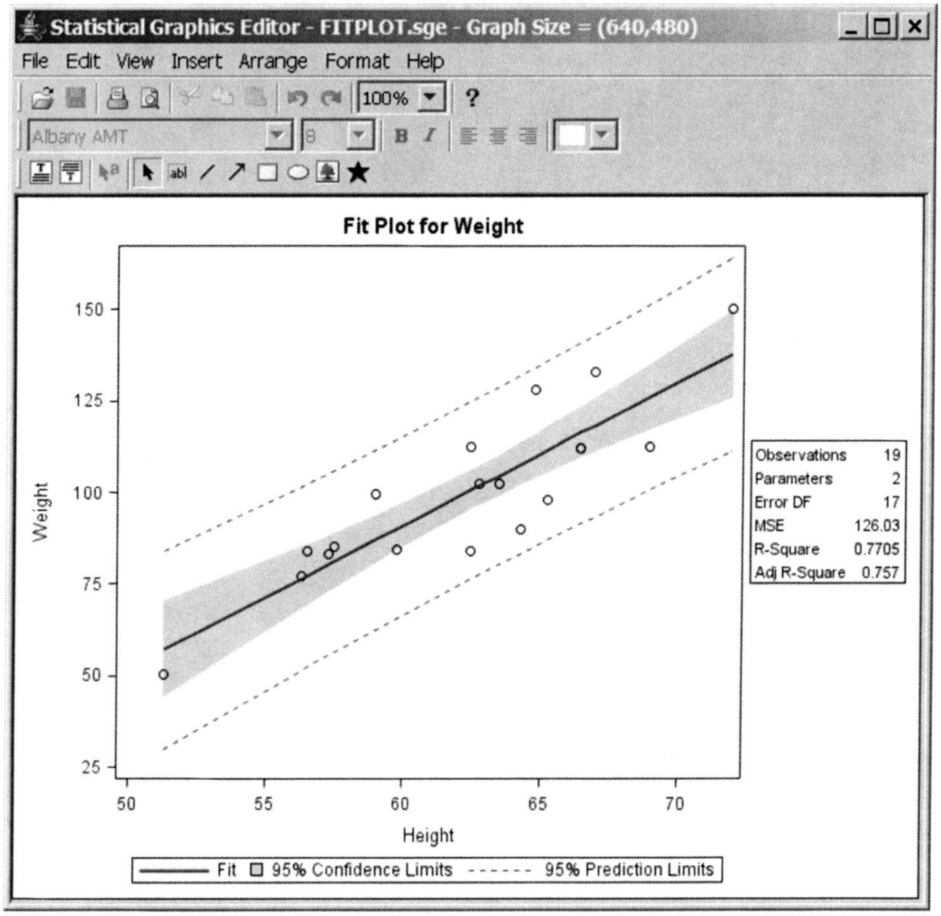

Figure 21.18 Point-and-Click Modifications Made with the ODS Graphics Editor

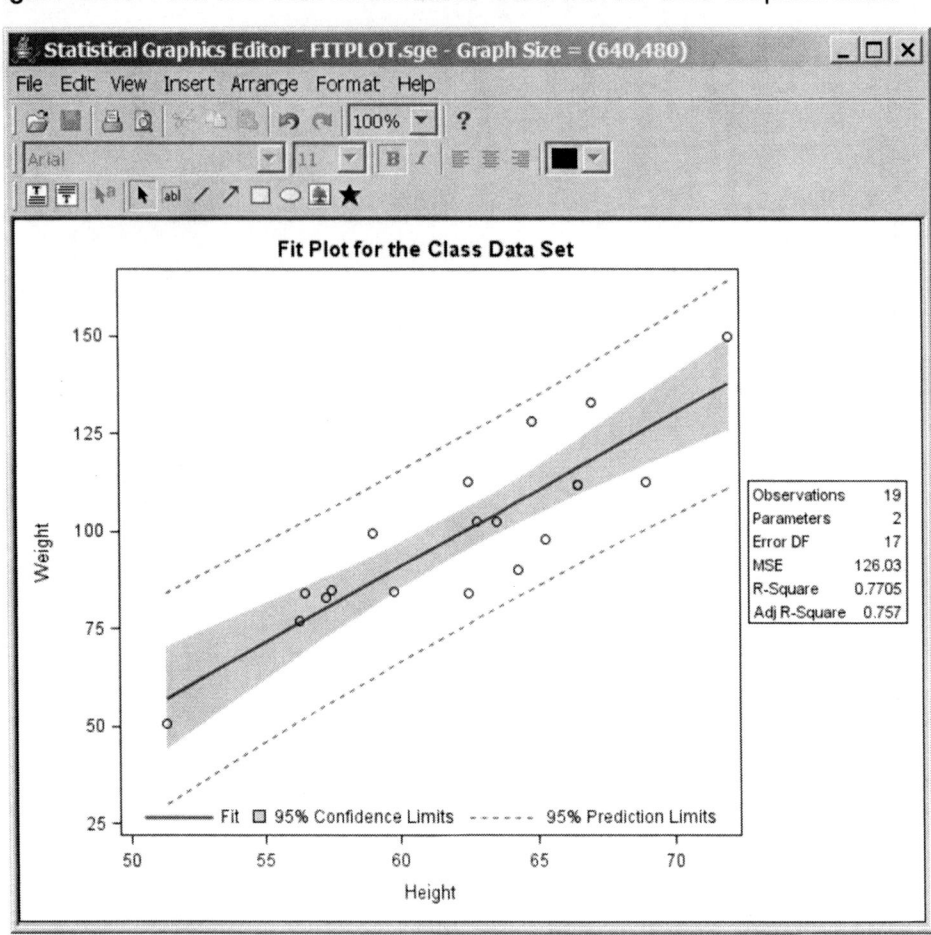

Procedures That Support ODS Graphics

SAS procedures that support ODS Graphics include the following:

SAS/STAT		**SAS/QC**	**SAS/ETS**
ANOVA	MIXED	ANOM	ARIMA
BOXPLOT	MULTTEST	CAPABILITY	AUTOREG
CALIS	NPAR1WAY	CUSUM	ENTROPY
CLUSTER	PHREG	MACONTROL	ESM
CORRESP	PLM	PARETO	EXPAND
FACTOR	PLS	RELIABILITY	MODEL
FREQ	PRINCOMP	SHEWHART	PANEL
GAM	PRINQUAL		SEVERITY
GENMOD	PROBIT	**Base**	SIMILARITY
GLIMMIX	QUANTREG	CORR	SYSLIN
GLM	REG	FREQ	TIMEID
GLMSELECT	ROBUSTREG	UNIVARIATE	TIMESERIES
KDE	RSREG		UCM
KRIGE2D	SEQDESIGN		VARMAX
LIFEREG	SEQTEST	**SAS/HPF**	X12
LIFETEST	SIM2D	HPF	
LOESS	SURVEYFREQ	HPFENGINE	
LOGISTIC	TPSPLINE		
MCMC	TRANSREG		
MDS	TTEST	**Risk**	
MI	VARIOGRAM	**Dimensions**	

For details about the specific graphs available with a particular procedure, see the PLOTS= option syntax and the "ODS Graphics" section in the corresponding procedure chapter. The procedure names in the preceding table are links to the "ODS Graphics" section for the SAS/STAT procedures.

Procedures That Support ODS Graphics and Traditional Graphics

A number of procedures that support ODS Graphics produced traditional graphics in previous releases of SAS. These include the UNIVARIATE procedure in Base SAS; the LIFEREG, LIFETEST, and REG procedures in SAS/STAT; and the ANOM, CAPABILITY, CUSUM, MACONTROL, PARETO, RELIABILITY, and SHEWHART procedures in SAS/QC. All of these procedures continue to produce traditional graphics, but in some cases, they do so only when ODS Graphics is not enabled. For more information about the interaction between traditional graphics and ODS graphics in other procedures, see the documentation for that procedure.

Traditional graphs are saved in SAS graphics catalogs and are controlled by the GOPTIONS statement. In contrast, ODS Graphics produces graphs in standard image file formats (not graphics catalogs), and their appearance and layout are controlled by ODS styles and templates.

Syntax

The following sections document some of the most commonly used options in the ODS GRAPHICS statement (section "ODS GRAPHICS Statement" on page 638) and other statements used with ODS Graphics (section "ODS Destination Statements" on page 641). You can find the complete syntax in the *SAS Output Delivery System: User's Guide*. In addition, information about the PLOTS= option is provided in the section "PLOTS= Option" on page 642. Statistical procedures that produce ODS Graphics all have a PLOTS= option that is used to select graphs and control some aspects of the graphs.

ODS GRAPHICS Statement

ODS GRAPHICS < OFF | ON > < / options > ;

The ODS GRAPHICS statement enables ODS to create graphics. By default, ODS Graphics is not enabled. You can enable ODS Graphics by using either of the following equivalent statements:

```
ods graphics on;
ods graphics;
```

You specify one of these statements prior to your procedure invocation, as illustrated in the examples beginning with "Default Plots for Simple Linear Regression with PROC REG" on page 609. Any procedure that supports ODS Graphics then produces graphics, either by default or when you specify procedure options for requesting particular graphs.

To disable ODS Graphics, specify the following statement:

```
ods graphics off;
```

The following is a subset of the options, syntax, and capabilities available in the ODS GRAPHICS statement. See the *SAS Output Delivery System: User's Guide* for more information.

ANTIALIAS=ON | OFF
> controls the use of antialiasing to smooth the components of a graph. Without antialiasing, pixels are simply set or not set. With antialiasing, pixels at the edge of a line or other object are set to an intermediate color, which makes smoother and more professional looking graphics. Text displayed in a graph is always antialiased. Antialiasing is very time consuming for larger graphical displays, and its benefits decrease as the number of points increases, so it is turned off by default for plots with many points. If the number of observations in the ODS output object exceeds the ANTIALIASMAX= threshold (10,000 by default), then antialiasing is not used, even if you specify the option ANTIALIAS=ON. The default is ANTIALIAS=ON.

ANTIALIASMAX=n

specifies the maximum number of markers or lines to be antialiased before antialiasing is disabled. For example, if there are more than 10,000 point markers and ANTIALIASMAX=10,000 (the default), then no markers are antialiased.

BORDER=ON | OFF

specifies whether to draw the graph with a border. BORDER=ON is the default.

HEIGHT=dimension

specifies the height of the graph. The default is HEIGHT=480PX (480 pixels). You can also specify height in inches (for example, HEIGHT=5IN) or centimeters (for example, HEIGHT=12CM).

IMAGEFMT=< image-file-type | **STATIC >**

specifies the image format for graphs. By default, IMAGEFMT=STATIC and ODS dynamically uses the best quality static image format for the active output destination. The available image formats include: BMP (Microsoft Windows device independent bitmap), DIB (Microsoft Windows device independent bitmap), EMF (Microsoft NT enhanced metafile), EPSI (Adobe encapsulated PostScript interchange), GIF (graphics interchange format), JFIF (JPEG file interchange format), JPEG (Joint Photographic Experts Group), PBM (portable bitmap), PCD (Photo CD), PDF (Portable Document Format), PICT (the QuickDraw Picture Format), PNG (Portable Network Graphic), PS (PostScript image file format), TIFF (tagged image file format), WMF (Microsoft Windows Metafile Format), XBM (X11 Bitmap graphics), and XPM (X11 Pixel Map graphics). If the specified image format is not valid for the active output destination, the device is automatically remapped to the default image format.

IMAGEMAP=ON | OFF

controls tooltip generation in the HTML destination. The default is IMAGEMAP=OFF, which means that no tooltips are generated. Tooltips are text boxes that appear in HTML output when you hover over a part of the plot with your mouse pointer (see Example 21.1).

IMAGENAME=< base-file-name **>**

specifies the base image file name. The default is the name of the output object. You can determine the name of the output object by using the ODS TRACE statement (see the section "Determining Graph Names and Labels" on page 647). The base image name should not include an extension. ODS automatically adds the increment value and the appropriate extension (which is specific to the output destination). See the section "Specifying Base File Names" on page 652 for an example.

LABELMAX=n

specifies the maximum number of labeled areas before labeling is disabled. For example, if LABELMAX=50, and there are more than 50 points with labels, then no points are labeled. The default is LABELMAX=200.

MAXLEGENDAREA=n

specifies the maximum percentage of the overall graph area that a legend can occupy. The default is MAXLEGENDAREA=20. Larger legends are dropped from the display.

RESET< =*option*>

resets one or more ODS GRAPHICS options to their default settings. The RESET and RESET=ALL options are equivalent. If you want to reset more than one option, but not all of the options, then you must specify RESET= separately for each option you reset (for example, `ods graphics on / reset=antialias reset=index;`). The RESET= options include the following:

ALL

resets all of the resettable options to their defaults.

ANTIALIAS

resets the ANTIALIAS= option to its default.

ANTIALIASMAX

resets the ANTIALIASMAX= option to its default.

BORDER

resets the BORDER= option to its default.

INDEX

resets the index counter that is appended to static image files.

HEIGHT

resets the HEIGHT= option to its default.

IMAGEMAP

resets the IMAGEMAP= option to its default.

LABELMAX

resets the LABELMAX= option to its default.

SCALE

resets the SCALE= option to its default.

TIPMAX

resets the TIPMAX= option to its default.

WIDTH

resets the WIDTH= option to its default.

SCALE=ON | OFF

specifies whether the fonts and symbol markers are scaled proportionally with the size of the graph. The default is SCALE=ON. For examples, see Figure 21.29 and Figure 21.30.

TIPMAX=*n*

specifies the maximum number of distinct tooltips permitted before tooltips are disabled. Tooltips are text boxes that appear when you hover over a part of the plot with your mouse pointer. For example, if TIPMAX=400, and there are more than 400 points in a scatter plot, then no tooltips appear. The default is TIPMAX=500.

WIDTH=dimension

specifies the width of the graph. The default is WIDTH=640PX (640 pixels). You can also specify widths in inches (for example, WIDTH=5in) or centimeters (for example, WIDTH=12cm).

ODS Destination Statements

ODS has a number of statements that control the destination of ODS output. The ODS destination statements that are most commonly used with ODS Graphics are: ODS DOCUMENT, ODS HTML, ODS LATEX, ODS LISTING, ODS PCL, ODS PDF, ODS PS, and ODS RTF. Specifying a statement opens a destination, unless the CLOSE option is specified. Each of the following statements opens an ODS destination:

```
ods html;
ods rtf;
ods html image_dpi=300;
ods listing style=Statistical;
```

Each of the following statements closes an ODS destination:

```
ods html close;
ods rtf close;
ods listing close;
```

There are two options that are commonly used in ODS destination statements to control aspects of ODS Graphics:

IMAGE_DPI=dpi

specifies the dots per inch (DPI), which is the image resolution for graphical output. The default varies depending on the destination. For example, the default is 100 for HTML and 200 for RTF.

STYLE=style-name

specifies the style definition. Commonly used styles include DEFAULT, LISTING, STATISTICAL, JOURNAL, JOURNAL2, RTF, and ANALYSIS.

Other options provide you with ways to control the files that are created. For example, the following statement opens the HTML destination:

```
ods html body='b.html' contents='c.html' frame='a.html';
```

This statement also writes the body of the output to the file *b.html*, the table of contents to the file *c.html*, and an overall frame containing both the contents and the output to the file *a.html*. Alternatively, you can specify FILE= instead of BODY=.

If you are using a destination for which individual graphs are created (for example, LISTING or HTML), you can use the GPATH= option to specify the directory where your graphics files are saved, as in the following example:

```
ods html gpath="C:\figures";
```

See the sections "Image File Types" on page 651, "Saving Graphics Image Files" on page 654, "LISTING Destination" on page 654, "HTML Destination" on page 654, and "LATEX Destination" on page 655 for more information about individual image files and options specified in the ODS Destination statements. For complete details about the ODS destination statements, see the *SAS Output Delivery System: User's Guide*.

PLOTS= Option

Each statistical procedure that produces ODS Graphics has a PLOTS= option that is used to select graphs and specify some options. The syntax of the PLOTS= option is as follows:

PLOTS < *(global-plot-options)* > < = *plot-request* < *(options)* > >
PLOTS < *(global-plot-options)* > < = (*plot-request* < *(options)* > < ... *plot-request* < *(options)* > >) >

The PLOTS= option has a common overall syntax for all statistical procedures, but the specific global plot options, plot requests, and plot options vary across procedures. This section discusses only a few of the options available in the PLOTS= option. For more information about the PLOTS= option, see the "Syntax" section for each procedure that produces ODS Graphics. There are only a limited number of things that you can control with the PLOTS= option. Most graphical details are controlled either by graph templates (see the section "Graph Templates" on page 687) or by styles (see the section "Styles" on page 664).

The PLOTS= option usually appears in the PROC statement. However, for some procedures, certain analyses and hence certain plots can appear only if an additional statement is specified. These procedures often have a PLOTS= option in that other statement. For example, the PHREG procedure has a PLOTS= option in the BAYES statement, which is used to perform a Bayesian analysis. See the "Syntax" section of each procedure chapter for more information. The following examples illustrate the syntax of the PLOTS= option:

```
plots=all
plots=none
plots=residuals
plots=residuals(smooth)
plots=(trace autocorr)
plots(unpack)
plots(unpack)=diagnostics
plots=diagnostics(unpack)
plots(only)=freqplot
plots=(scree(unpack) loadings(plotref) preloadings(flip))
plots(unpack maxparmlabel=0 stepaxis=number)=coefficients
plots(sigonly)=(rawprob adjusted(unpack))
```

Also see the "Getting Started" sections "Survival Estimate Plot with PROC LIFETEST" on page 612, "Contour and Surface Plots with PROC KDE" on page 613, "Contour Plots with PROC KRIGE2D" on page 615, "LS-Means Diffogram with PROC GLIMMIX" on page 621, and "Principal Component Analysis Plots with PROC PRINCOMP" on page 623 for examples of the PLOTS= option.

The simplest PLOTS= specifications are of the form PLOTS=*plot-request* or PLOTS=(*plot-requests*). When there is more than one plot request, the plot-request list must appear in parentheses. Each plot request either requests a plot (for example, RESIDUALS) or provides you with a place to specify plot-specific options (for example, DIAGNOSTICS(UNPACK)). Some simple and typical plot requests are explained next:

- PLOTS=ALL requests all plots that are relevant to the analysis. This does not mean that all plots that the procedure can produce are produced. Plots that are produced for one set of options might not appear with PLOTS=ALL and a different set of options. In some cases, certain plots are not produced unless certain options or statements outside the PLOTS= option are specified.

- PLOTS=NONE disables ODS Graphics for just that step. You can use this option instead of specifying ODS GRAPHICS OFF before a procedure step and ODS GRAPHICS ON after the step when you want to suppress graphics for only that step.

- PLOTS=RESIDUALS requests a plot of residuals in a modeling procedure such as PROC REG.

- PLOTS=RESIDUALS(SMOOTH) requests the residuals plot along with a smooth fit function.

- PLOTS=(TRACE AUTOCORR) requests trace and autocorrelation plots in procedures with Bayesian analysis options.

Global plot options appear in parentheses after the option name and before the equal sign. These options affect many or all of the plots. The UNPACK option is a commonly used global plot option. It specifies that plots that are normally produced with multiple plots per panel (or "packed") should be unpacked and appear in multiple panels with one plot in each panel. The specification PLOTS(UNPACK)=(*plot-requests*) unpacks all paneled plots. The UNPACK option is also used as an option in a plot request when you only want to unpack certain panels. For example, the option PLOTS=(DIAGNOSTICS(UNPACK) PARTIAL PREDICTIONS) unpacks just the diagnostics panel. In some cases, unpacked plots contain additional information that is not found in the smaller packed versions. The UNPACK option is not available for all plot requests; it is just available with plots that have multiple panels by default.

Another commonly used global plot option is the ONLY option. Many procedures produce default plots, and additional plots can be requested in the PLOTS= option. Specifying PLOTS=(*plot-requests*) while omitting the default plots does not prevent the default plots from being produced. The ONLY option is used when you only want to see the plots specifically listed in the plot-request list. Procedures that produce no default plots typically do not provide an ONLY option. You can use ODS SELECT and ODS EXCLUDE (see the section "Selecting and Excluding Graphs" on page 649) to select and exclude graphs, but in some situations the ONLY option is more convenient. It is typically more efficient to select plots by using the PLOTS(ONLY)= option, because the procedure does not do extra work to generate a plot that is excluded by the PLOTS(ONLY)= option. In contrast, ODS SELECT and ODS EXCLUDE have their effect after the procedure has done the work to generate the plot.

Selecting and Viewing Graphs

This section describes techniques for selecting and viewing your graphs. Topics include:

- specifying an ODS destination for graphics
- viewing your graphs in the SAS windowing environment
- referring to graphs by name when using ODS
- selecting and excluding graphs from your output

Specifying an ODS Destination for Graphics

If you do not specify an ODS destination, then the LISTING destination is used by default. Here is an example of how you can specify the HTML destination:

```
ods graphics on;
ods html;

proc reg data=sashelp.class;
   model Weight = Height;
run; quit;

ods html close;
```

This ODS HTML statement creates an HTML file with a default name. See the section "Specifying a File for ODS Output" on page 645 to see how to specify a file name. Other destinations are specified in a similar way. For example, you can specify an RTF destination with the following statements:

```
ods graphics on;
ods rtf;

. . .

ods rtf close;
```

The destinations that ODS supports for graphics are as follows:

Destination	Destination Family
DOCUMENT	
HTML	MARKUP
LATEX	MARKUP
LISTING	
PCL	PRINTER
PDF	PRINTER
PS	PRINTER
RTF	

You can close the LISTING destination if you are only interested in displaying your output in a different destination. For example, if you want to see your output only in the RTF destination, you can specify the following statements:

```
ods graphics on;
ods listing close;
ods rtf;
```

. . .

```
ods rtf close;
ods listing;
```

Closing unneeded destinations makes your jobs run faster and creates fewer files. More generally, it makes your jobs consume fewer resources, because a graph is otherwise created for every open destination. The last statement opens the LISTING destination after you are finished using the RTF destination.

You can also use the ODS OUTPUT destination to create an output data set from the data object used to make a plot. Here is an example:

```
ods graphics on;

proc reg data=sashelp.class;
   ods output fitplot=myfitplot;
   model Weight = Height;
run; quit;
```

Specifying a File for ODS Output

You can specify a file name for your output with the FILE= option in the ODS destination statement, as in the following example:

```
ods html file="test.htm";
```

The output is written to the file *test.htm*, which is saved in the SAS current folder. At startup, the SAS current folder is the same directory in which you started your SAS session. If you are using the SAS windowing environment, then the current folder is displayed in the status line at the bottom of the main SAS window. If you do not specify a file name for your output, then the SAS System provides a default file name, which depends on the ODS destination. This file is saved in the SAS current folder. You can always check the SAS log to verify the name of the file in which your output is saved. For example, suppose you specify the following statement:

```
ods html;
```

Then the following message is displayed in the SAS log:

```
NOTE: Writing HTML Body file: sashtml.htm
```

The default file names for each destination are specified in the SAS Registry. For example, Figure 21.31 shows that the default file name in the SAS Registry for the RTF destination is *sasrtf.rtf*. For more information, see the SAS Companion for your operating system.

Viewing Your Graphs in the SAS Windowing Environment

The mechanism for viewing graphics created with ODS can vary depending on your operating system, which viewers are installed on your computer, and the ODS destination you have selected. If you do not specify an ODS destination, then the LISTING destination is used by default. If you are using the SAS windowing environment, go to the Results window and find the icon for the corresponding graph. You can double-click the graph icon to display the graph in the default viewer that is configured on your computer for the corresponding image file type (see Figure 21.16).

If you are using the SAS windowing environment and you specify an HTML destination, then the results are displayed by default in the SAS Results window as they are being generated. Depending on your configuration, this can also apply to the PDF and RTF destinations. If you are using the LATEX or the PS destinations, you must use a PostScript viewer, such as GSview. For information about the windowing environment in a different operating system, see the SAS Companion for that operating system.

If you do not want to view the results as they are being generated, then select **Tools ▶ Options ▶ Preference** from the menu at the top of the main SAS window. Then in the **Results** tab, clear **View results as they are generated** checkbox.

You can change the default to use an external viewer instead of the Results window. Select **Tools ▶ Options ▶ Preferences** from the menu at the top of the main SAS window. Then in the **Results** tab select **Preferred web browser**. Your results are displayed in the default viewer that is configured on your computer for the corresponding destination.

You can also choose your browser for HTML output. Select **Tools ▶ Options ▶ Preferences** from the menu at the top of the main SAS window. Then in the **Web** tab, select **Other browser**, and type (or browse) the path of your preferred browser.

Determining Graph Names and Labels

Procedures assign a name to each graph they create with ODS Graphics. This enables you to refer to ODS graphs in the same way that you refer to ODS tables (see the section "The ODS Statement" on page 541 in Chapter 20, "Using the Output Delivery System"). You can determine the names of graphs in several ways:

- You can look up graph names in the "ODS Graphics" section of chapters for procedures that use ODS Graphics. For example, see the section "ODS Graphics" on page 6301 in Chapter 74, "The REG Procedure."

- You can use the Results window to view the names of ODS graphs created in your SAS session. See the section "ODS and the SAS Results Window" on page 546 in Chapter 20, "Using the Output Delivery System," for more information.

- You can use the ODS TRACE ON statement to list the names of graphs created by your SAS session. This statement adds identifying information in the SAS log (or optionally in the SAS LISTING) for each graph that is produced. See the section "The ODS Statement" on page 541 in Chapter 20, "Using the Output Delivery System," for more information.

The graph name is not the same as the name of the image file that contains the graph (see the section "Naming Graphics Image Files" on page 652).

This example revisits the analysis described in the section "Contour and Surface Plots with PROC KDE" on page 613. To determine which output objects are created by ODS, you specify the ODS TRACE ON statement prior to the procedure statements as follows:

```
ods graphics on;
ods trace on;

proc kde data=bivnormal;
   bivar x y / plots=contour surface;
run;

ods trace off;
```

The trace record from the SAS log is as follows:

```
Output Added:
-------------
Name:       Inputs
Template:   Stat.KDE.Inputs
Path:       KDE.Bivar1.x_y.Inputs
-------------
```

```
Output Added:
--------------
Name:       Controls
Template:   Stat.KDE.Controls
Path:       KDE.Bivar1.x_y.Controls
--------------

Output Added:
--------------
Name:       ContourPlot
Label:      Contour Plot
Template:   Stat.KDE.Graphics.Contour
Path:       KDE.Bivar1.x_y.ContourPlot
--------------

Output Added:
--------------
Name:       SurfacePlot
Label:      Density Surface
Template:   Stat.KDE.Graphics.Surface
Path:       KDE.Bivar1.x_y.SurfacePlot
--------------
```

By default, PROC KDE creates table objects named **Inputs** and **Controls**, and it creates graph objects named **ContourPlot** and **SurfacePlot**. In addition to the name, the trace record provides the label, template, and path for each output object. Graph templates are distinguished from table templates by a naming convention that uses the procedure name in the second level and **Graphics** in the third level. For example, the fully qualified template name for the surface plot created by PROC KDE is **Stat.KDE.Graphics.SurfacePlot**.

You can specify the LISTING option in the ODS TRACE ON statement to write the trace record to the LISTING destination as follows:

```
ods trace on / listing;
```

Each table and graph has a path (or name path), which was previously shown in the trace output. The path consists of the plot name preceded by the names of one or more output groups. Each table and graph also has a label path, which can be seen by adding the LABEL option to the ODS TRACE ON statement, after a slash, as follows:

```
ods trace on / label;

proc kde data=bivnormal;
   bivar x y / plots=contour surface;
run;

ods trace off;
```

A portion of the trace output is shown next:

```
Path:          KDE.Bivar1.x_y.Inputs
Label Path:    'The KDE Procedure'.'Bivariate Analysis'.'x and y'.'KDE.Bivar1.x_y'

Path:          KDE.Bivar1.x_y.Controls
Label Path:    'The KDE Procedure'.'Bivariate Analysis'.'x and y'.'KDE.Bivar1.x_y'

Path:          KDE.Bivar1.x_y.ContourPlot
Label Path:    'The KDE Procedure'.'Bivariate Analysis'.'x and y'.'Contour Plot'

Path:          KDE.Bivar1.x_y.SurfacePlot
Label Path:    'The KDE Procedure'.'Bivariate Analysis'.'x and y'.'Density Surface'
```

The label path contains the information that you see in the HTML table of contents. Names are fixed, they do not vary, and they are not data- or context-dependent. In contrast, labels often reflect data- or context-dependent information.

Selecting and Excluding Graphs

You can use the ODS SELECT and ODS EXCLUDE statements along with graph and table names to specify which ODS outputs are displayed. See the section "The ODS Statement" on page 541 in Chapter 20, "Using the Output Delivery System," for more information about how to use these statements.

This section shows several examples of selecting and excluding graphs by using the data set and trace output created in the section "Determining Graph Names and Labels" on page 647. The following statements use the ODS SELECT statement to select only two graphs, `ContourPlot` and `SurfacePlot`, for display in the output:

```
proc kde data=bivnormal;
   ods select ContourPlot SurfacePlot;
   bivar x y / plots=contour surface;
run;
```

Equivalently, the following statements use the ODS EXCLUDE statement to exclude the two tables:

```
proc kde data=bivnormal;
   ods exclude Inputs Controls;
   bivar x y / plots=contour surface;
run;
```

You can select or exclude graphs by using either the name or the label. Labels must be specified in quotes. In the context of this example, the following two statements are equivalent:

```
ods select contourplot;
ods select 'Contour Plot';
```

You can also specify multiple levels of the path, as in the following example:

```
ods select x_y.contourplot;
ods select 'x and y'.'Contour Plot';
ods select 'x and y'.contourplot;
ods select x_y.'Contour Plot';
```

Name and label paths can be mixed, as in the last two statements. All four of the preceding statements select the same plot. Furthermore, selection based directly on the names and labels is case insensitive. The following all select the same plot:

```
ods select x_y.contourplot;
ods select 'x and y'.'Contour Plot';
ods select X_Y.CONTOURPLOT;
ods select 'X AND Y'.'CONTOUR PLOT';
```

It is sometimes useful to specify a WHERE clause in an ODS SELECT or ODS EXCLUDE statement. This enables you to specify expressions based on either the name path or the label path. You can base your selection on two automatic variables _path_ and _label_. The following two statements select every object whose path contains the string 'plot' and every object whose label path contains the string 'plot', respectively, ignoring the case in the name and label:

```
ods select where = (lowcase(_path_)  ? 'plot');
ods select where = (lowcase(_label_) ? 'plot');
```

The question mark operator means that the second expression (the string 'plot') is contained in the first expression (the lowercase version of the name or label). For example, all of the following names match 'plot' in the WHERE clause: plot, SurfacePlot, SURFACEPLOT, FitPlot, pLoTtInG, Splotch, and so on. Since WHERE clause selection is based on SAS string comparisons, selection is case sensitive. The LOWCASE function is used to ensure a match even when the specified string does not match the case of the actual name or label.

WHERE clauses are particularly useful when you want to select all of the objects in a group. A group is a level of the name path or label path hierarchy before the last level. In the following step, all of the objects whose name path contains 'DiagnosticPlots' are selected:

```
proc reg data=sashelp.class plots(unpack);
   ods select where = (_path_ ? 'DiagnosticPlots');
   model Weight = Height;
run; quit;
```

These are the plots that come from unpacking the PROC REG diagnostics panel of plots. All are in a group named 'DiagnosticPlots'.

Graphics Image Files

Accessing your graphs as individual image files is useful when you want to include them in various types of documents. The default image file type depends on the ODS destination, but there are other supported image file types that you can specify. You can also specify the names for your graphics image files and the directory in which you want to save them. This section describes the image file types supported by ODS Graphics, and it explains how to name and save graphics image files.

Image File Types

If you are using the LISTING, HTML or LATEX destinations, your graphs are individually produced in a specific image file type, such as PNG (Portable Network Graphics). If you are using a destination in the PRINTER family or the RTF destination, the graphs are contained in the ODS output file and cannot be accessed as individual image files. However, you can open an RTF output file in Microsoft Word and then copy and paste the graphs into another document, such as a Microsoft PowerPoint presentation. This is illustrated in Example 21.2.

Table 21.1 shows the various ODS destinations supported by ODS Graphics, the viewer that is appropriate for displaying graphs in each destination, and the image file types supported for each destination.

Table 21.1 Destinations and Image File Types Supported by ODS Graphics

Destination	Destination Family	Recommended Viewer	Image File Types
DOCUMENT		Not applicable	Not applicable
HTML	MARKUP	Web browser	PNG (default), GIF, JPEG,
LATEX	MARKUP	PostScript or PDF viewer after compiling the LaTeX file	PostScript (default), EPSI, GIF, JPEG, PDF, PNG
LISTING		Default viewer in your system for the specified file type	PNG (default), GIF, BMP, DIB, EMF, EPSI, GIF, JFIF, JPEG, PBM, PS, TIFF, WMF
PCL	PRINTER	not applicable	Contained in PRN file
PDF	PRINTER	PDF viewer, such as Adobe Reader	Contained in PDF file
PS	PRINTER	PostScript viewer, such as GSview	Contained in PostScript file
RTF		Word processor, such as Microsoft Word	Contained in RTF file

For destinations such as PDF and RTF, you can control the types of the images that are contained in the file even though individual files are not made for each image. The default image file type is PNG, and other image types are available. See the *SAS Output Delivery System: User's Guide* for more information.

Naming Graphics Image Files

The following discussion applies to the destinations where ODS graphs are created as individual image files (for example, HTML, LISTING, and LATEX). The names of graphics image files are determined by a base file name, an index counter, and an extension. By default, the base file name is the ODS graph name (see the section "Determining Graph Names and Labels" on page 647). There is an index counter for each base file name. The extension indicates the image file type. The first time a graph object with a given base file name is created, the file name consists only of the base file name and the extension. If a graph with the same base file name is created multiple times, then an index counter is appended to the base file name to avoid overwriting previously created images.

To illustrate, consider the following statements:

```
proc kde data=bivnormal;
   ods select ContourPlot SurfacePlot;
   bivar x y / plots=contour surface;
run;
```

If you run this step at the beginning of a SAS session, the two graphics image files created are *ContourPlot.png* and *SurfacePlot.png*. If you immediately rerun these statements, then ODS creates the same graphs in different image files named *ContourPlot1.png* and *SurfacePlot1.png*. The next time, the image files are named *ContourPlot2.png* and *SurfacePlot2.png*. The index starts at zero, and one is added each time the same name is used. Note, however, that when the index is at zero, it is not added to the file name.

Resetting the Index Counter

You can specify the RESET=INDEX option in the ODS GRAPHICS statement to reset the index counter. This is useful when you need to have predictable names. It is particularly useful when you are running a SAS program multiple times in the same session. The following statement resets the index:

```
ods graphics on / reset=index;
```

The index counter is reinitialized at the beginning of your SAS session or if you specify the RESET=INDEX option in the ODS GRAPHICS statement. Graphics image files with the same name are overwritten.

Specifying Base File Names

You can specify a base file name for all your graphics image files with the IMAGENAME= option in the ODS GRAPHICS statement as follows:

```
ods graphics on / imagename="MyName";
```

You can also specify the RESET=INDEX option as follows:

```
ods graphics on / reset=index imagename="MyName";
```

The IMAGENAME= option overrides the default base file name. With the preceding statement, the graphics image files are named *MyName*, *MyName1*, *MyName2*, and so on.

Specifying Image File Types

You can specify the image file type for the LISTING, HTML, or LATEX destinations with the IMAGEFMT= option in the ODS GRAPHICS statement as follows:

```
ods graphics on / imagefmt=gif;
```

For more information, see the section "ODS GRAPHICS Statement" on page 638.

Naming Graphics Image Files with Multiple Destinations

Since the index counter depends only on the base file name, if you specify multiple ODS destinations for your output, then the index counter is increased independently of the destination. For example, the following statements create image files named *ContourPlot.png* and *SurfacePlot.png* that correspond to the LISTING destination; and *ContourPlot1.png* and *SurfacePlot1.png* that correspond to the HTML destination:

```
ods listing;
ods html;
ods graphics on / reset;

proc kde data=bivnormal;
   ods select ContourPlot SurfacePlot;
   bivar x y / plots=contour surface;
run;

ods _all_ close;
ods listing;
```

When you specify one of the destinations in the PRINTER family or the RTF destination, your ODS graphs are embedded in the document, so the index counter is not affected. For example, the following statements create image files *ContourPlot.png* and *SurfacePlot.png* for the LISTING destinations, but no image files for the RTF destination:

```
ods listing;
ods rtf;
ods graphics on / reset;

proc kde data=bivnormal;
   ods select ContourPlot SurfacePlot;
   bivar x y / plots=contour surface;
run;

ods _all_ close;
```

Saving Graphics Image Files

Knowing where your graphics image files are saved and how they are named is particularly important if you are running in batch mode, if you have disabled the SAS Results window (see the section "Viewing Your Graphs in the SAS Windowing Environment" on page 646), or if you plan to access the files for inclusion in a paper or presentation. The following discussion assumes you are running SAS under the Windows operating system. If you are running on a different operating system, see the SAS Companion for your operating system.

Your graphics image files are saved by default in the SAS current folder. If you are using the SAS windowing environment, the current folder is displayed in the status line at the bottom of the main SAS window. If you are running your SAS programs in batch mode, the graphs are saved by default in the same directory where you started your SAS session. For example, suppose the SAS current folder is *C:\myfiles*. If you specify the ODS GRAPHICS statement, then your graphics image files are saved in the directory *C:\myfiles*. Unlike traditional graphics, ODS Graphics are not saved in a catalog in your WORK directory.

With the LISTING, HTML, and LATEX destinations, you can specify a directory for saving your graphics image files. With a destination in the PRINTER family and with the RTF destination, you can specify a directory only for your output file. The remainder of this discussion provides details for each destination type.

LISTING Destination

If you are using the LISTING destination, the individual graphs are created as PNG files by default. You can use the GPATH= option in the ODS LISTING statement to specify the directory where your graphics files are saved. For example, if you want to save your graphics image files in *C:\figures*, then you can specify the following:

```
ods listing gpath="C:\figures";
```

It is important to note that the GPATH= option applies only to ODS Graphics. It does not affect the behavior of graphics created with traditional SAS/GRAPH procedures.

HTML Destination

If you are using the HTML destination, the individual graphs are created as PNG files by default. You can use the PATH= and GPATH= options in the ODS HTML statement to specify the directory where your HTML and graphics files are saved, respectively. This also gives you more control over your graphs. For example, if you want to save your HTML file named *test.htm* in the *C:\myfiles* directory, but you want to save your graphics image files in *C:\myfiles\png*, then you can specify the following:

```
ods html path  ="C:\myfiles"
         gpath = "C:\myfiles\png"
         file  = "test.htm";
```

When you specify the URL= suboption with the GPATH= option, SAS creates relative paths for the links and references to the graphics image files in the HTML file. This is useful for building output files that are easily moved from one location to another. For example, the following statements create a relative path to the *png* directory in all the links and references contained in *test.htm*:

```
ods html path  = "C:\myfiles"
         gpath = "C:\myfiles\png" (url="png/")
         file  = "test.htm";
```

If you do not specify the URL= suboption, SAS creates absolute paths that are hard-coded in the HTML file. These can cause broken links if you move the files. For more information, see the ODS HTML statement in the *SAS Output Delivery System: User's Guide*.

LATEX Destination

LaTeX is a document preparation system for high-quality typesetting. The ODS LATEX statement produces output in the form of a LaTeX source file that is ready to compile in LaTeX. When you request ODS Graphics for a LATEX destination, ODS creates the requested graphs as PostScript files by default, and the LaTeX source file includes references to these image graphics files. You can compile the LaTeX file, or you can ignore this file and simply access the individual PostScript files to include your graphs in a different LaTeX document, such as a paper that you are writing. You can specify the PATH= and GPATH= options in the ODS LATEX statement, as explained previously for the ODS HTML statement. See Example 21.3 for an illustration. The ODS LATEX statement is an alias for the ODS MARKUP statement with the TAGSET=LATEX option. For more information, see the *SAS Output Delivery System: User's Guide*.

The default image file type for the LATEX destination is PostScript. When you use LaTeX to compile your document, the graphics format for included images is Postscript. However, if you prefer to use pdfLaTeX, you can specify a different format such as JPEG, PDF, or PNG, any of which can be directly included into your pdfLaTeXdocument. To specify one of these formats, you use the IMAGEFMT= option in the ODS GRAPHICS statement. For more information, see the LaTeX documentation for the `graphicx` package.

Creating Graphs in Multiple Destinations

This section illustrates how to send your output to more than one destination with a single execution of your SAS statements. For example, to create the default LISTING output and also both HTML and RTF output, you can specify the ODS HTML and the ODS RTF statements before your procedure statements.

```
ods html;
ods rtf;

   . . .

ods _all_ close;
```

The ODS _ALL_ CLOSE statement closes all open destinations.

You can also specify multiple instances of the same destination. For example, using the data in the section "Contour and Surface Plots with PROC KDE" on page 613, the following statements save the contour plot to the file *contour.pdf* and the surface plot to the file *surface.pdf*:

```
ods pdf file="contour.pdf";
ods pdf select ContourPlot;
ods pdf(id=srf) file="surface.pdf";
ods pdf(id=srf) select SurfacePlot;
ods graphics on;

proc kde data=bivnormal;
   ods select ContourPlot SurfacePlot;
   bivar x y / plots=contour surface;
run;

ods _all_ close;
```

The ID= option assigns the name `srf` to the second instance of the PDF destination. Without the ID= option, the second ODS PDF statement closes the destination that was opened by the previous ODS PDF statement, and it opens a new instance of the PDF destination. In that case, the file *contour.pdf* is not created. For more information, see the ODS PDF statement in the *SAS Output Delivery System: User's Guide*.

Graph Size and Resolution

ODS provides options for specifying the size and resolution of graphs. You can specify the size of a graph in the ODS GRAPHICS statement and the resolution in an ODS destination statement. There are two other ways to change the size of a graph, but they are rarely needed. The three methods are as follows:

- Usually, you specify WIDTH=, HEIGHT=, or both in the ODS GRAPHICS statement to change the size of a graph.

- To modify the size of a particular graph, specify the dimensions with the DESIGNHEIGHT= and DESIGNWIDTH= options in the BEGINGRAPH statement in the template definition. Some templates contain the specification DESIGNWIDTH=DEFAULTDESIGNHEIGHT, which sets the width of the graph to the default height, or DESIGNHEIGHT=DEFAULTDESIGNWIDTH, which sets the height of the graph to the default width.

- To modify the size of all of your ODS graphs, specify the dimensions with the OUTPUTHEIGHT= and OUTPUTWIDTH= options in the style definition.

The following examples show typical ways to change the size of your graphs:

```
ods graphics on / width=6in;
ods graphics on / height=4in;
ods graphics on / width=4.5in height=3.5in;
```

The dimensions of the graph can be specified in pixels (for example, 200PX), inches (for example, 3IN), or centimeters (for example, 8CM). The default dimensions of ODS Graphics are 640 pixels wide and 480 pixels high, and these values determine the default aspect ratio. The actual size of the graph in inches depends on your printer or display device. For example, if the resolution of your printer is 100 dots per inch, and you want a graph that is 4 inches wide, you should set the width to 400 pixels.

If you specify only one dimension, the other is determined by the default aspect ratio—that is, height=0.75 × width. For best results, you should create your graphs by using the exact size that is used to display the graphs in your paper or presentation. In other words, avoid generating them at one size and then expanding or shrinking them for inclusion into the your document.

By default, fonts and symbol markers are automatically scaled with the size of the graph. You can suppress this scaling with the SCALE= option, as in the following example:

```
ods graphics on / scale=off;
```

The default resolution of graphs created with HTML and LISTING is 100 DPI (dots per inch), whereas the default with RTF is 200 DPI. The 200 DPI value is recommended if you are copying and pasting graphs into a Microsoft PowerPoint presentation or a Microsoft Word document. Graphs shown in SAS/STAT documentation are typically generated at 300 DPI for display in PDF and 100 DPI for display in HTML.

You can change the resolution with the IMAGE_DPI= option in any ODS destination statement, as in the following example:

```
ods html image_dpi=300;
```

An increase in resolution often improves the quality of the graphs, but it also greatly increases the size of the image file. Going from 100 DPI to 300 DPI increases the size of the image file by roughly a factor of $(300/100)^2 = 9$. Even when you are using a higher DPI for most of your graphs, you should consider using a lower DPI for some, such as contour plots, that create large files even at a lower DPI.

If you increase the resolution, you might need to compensate by reducing the size of the graph, as in the following example:

```
ods graphics on / width=4.5in  height=3.5in;
```

Increasing DPI also increases the amount of memory needed for your program to complete. You can increase the amount of memory available to ODS Graphics with an option when you invoke SAS, as in the following example:

```
-jreoptions '(-Xmx256m)'
```

You can modify the default amount of memory available to ODS Graphics by changing JREOPTIONS in your SAS configuration file to the settings –Xmx*nnn*m –Xms*nnn*m, where *nnn* is the amount

658 ✦ *Chapter 21: Statistical Graphics Using ODS*

of memory in megabytes. An example is –Xmx256m –Xms256m. In either case, the exact syntax varies depending on your operating system, and the amount of memory that you can allocate varies from system to system. For more information, see the SAS Companion for your operating system.

ODS Graphics Editor

The ODS Graphics Editor is a point-and-click interface that you can use to modify a specific graph created by ODS Graphics. For example, if you need to enhance a graph for a paper or presentation, you can use the ODS Graphics Editor to customize the title, modify the axis labels, annotate particular data points, and change graph element properties such as fonts, colors, and line styles.

This section explains how to enable ODS Graphics to create editable graphs and how to invoke the ODS Graphics Editor. You can use the ODS Graphics Editor in the SAS windowing environment, provided that the LISTING destination is open and that you have first enabled ODS Graphics to create editable graphs. **NOTE:** The LISTING destination is typically open by default. There are three steps you must take to edit a graph:

1 You must first enable the creation of editable graphs in one of three ways:
- use an ODS statement to temporarily enable this feature
- use a SAS command to temporarily enable this feature
- use the SAS Registry Editor to permanently enable this feature

Creating editable graphs takes additional resources, so you might not want to permanently enable this feature.

2 You submit your SAS code and create editable graphs.

3 You invoke the ODS Graphics Editor and edit the plot.

Step 2 involves submitting SAS code in the usual way, and no special instructions are needed for creating graphs that can be edited. Steps 1 and 3 are explained in more detail in the following sections.

Enabling the Creation of Editable Graphs

Temporarily Enable Creation of Editable Graphs by Using an ODS Statement

You can enable the creation of editable graphs within a SAS session by submitting the following statement:

```
ods listing sge=on;
```

You can disable the creation of editable graphs by submitting the following statement:

```
ods listing sge=off;
```

Temporarily Enable Creation of Editable Graphs by Using a SAS Command

Alternatively, you can enable the creation of editable graphs for the duration of your SAS session by first selecting the Results window and then entering `sgedit on` in the command line. SAS confirms that the creation of editable graphs is enabled by displaying a message in the bottom left corner of the SAS window. The command must be entered from the Results window. If you enter it from any other window, it is ignored.

Permanently Enable Creation of Editable Graphs across SAS Sessions

You can create a default setting that enables or disables the creation of editable graphs across SAS sessions via the 'ODS Graphics Editor' setting in the SAS Registry. You can change this setting in the SAS windowing environment as follows:

1 Open the Registry Editor by entering **regedit** in the command line.

2 Select **SAS_REGISTRY ► ODS ► GUI ► RESULTS**.

3 In the **Value Data** field, click **ODS Graphics Editor** to open the Edit String Value window, and type **On** to enable the creation of editable graphs or type **Off** to disable it.

4 Click **OK**.

Editing a Graph with the ODS Graphics Editor

The ODS Graphics Editor is illustrated using the following example:

```
data sasuser.growth;
   input country $ GDP LFG EQP NEQ GAP @@;
   datalines;
Argentin  0.0089 0.0118 0.0214 0.2286 0.6079
Austria   0.0332 0.0014 0.0991 0.1349 0.5809

  ... more lines ...

Zambia   -0.0110 0.0275 0.0702 0.2012 0.8695
Zimbabwe  0.0110 0.0309 0.0843 0.1257 0.8875
;

ods graphics on;
ods listing style=statistical sge=on;
```

```
proc robustreg data=sasuser.growth
             plots=(ddplot histogram);
  model GDP  = LFG GAP EQP NEQ / diagnostics leverage;
  output out=robout r=resid sr=stdres;
run;

ods listing sge=off;
```

The DATA and the PROC ROBUSTREG steps are submitted to SAS, in this case from the SAS windowing environment, as shown in Figure 21.19. Two versions of the graph are created: one in an uneditable PNG file (for example, *DDPlot.png*) and one in an editable SGE file (for example, *DDPlot.sge*). Both are saved in the SAS current folder. You can edit the graph in one of three ways:

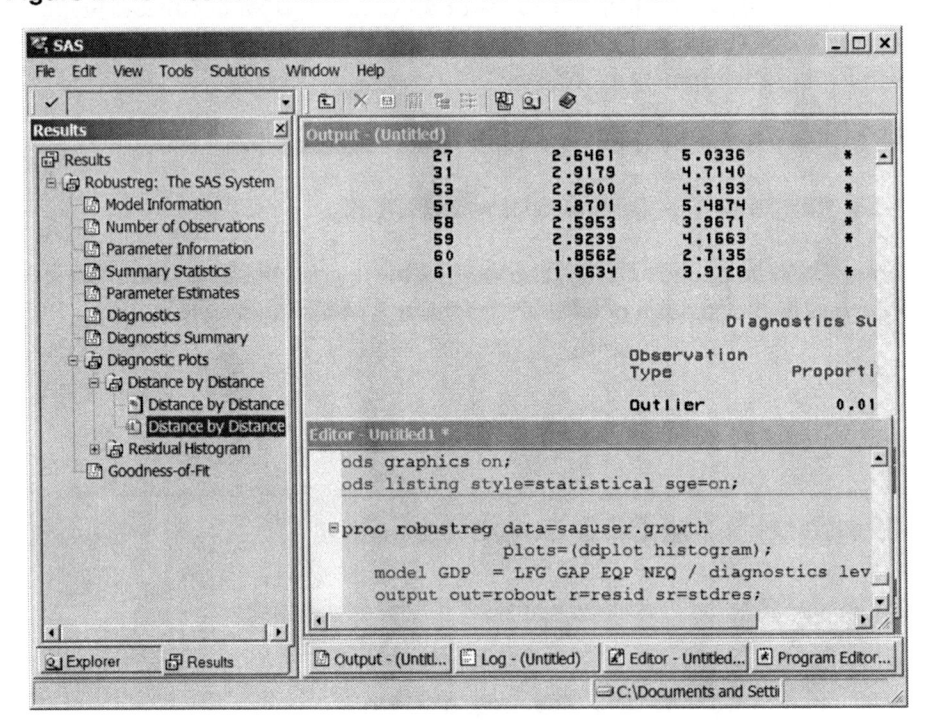

Figure 21.19 Results Window with Icons for Editable Plots

- In the Results window, double-click the second graph icon for the graph you want to edit (see Figure 21.19). The second icon corresponds to the SGE file, and the first icon corresponds to the PNG file. Clicking the first graph icon invokes a host-dependent graph viewer (for example, Microsoft Photo Editor on Windows), not the ODS Graphics Editor. **NOTE:** The Editor window might be hidden behind other windows in the SAS windowing environment.

- You can edit the graph by selecting it in the SAS Explorer window. You must first navigate to the SAS current folder and to the SGE files.

- You can open the graph from outside of SAS. For example, if you are running SAS under the Windows operating system, you can click on the graph's SGE file to open it with the ODS Graphics Editor.

Figure 21.20 shows the ODS Graphics Editor window for the editable diagnostic plot created by PROC ROBUSTREG. In Figure 21.21, various tools in the ODS Graphics Editor are used to modify the title and annotate a particular point. The edited plot can be saved as a PNG file or as an SGE file by selecting **File ▶ Save As**. After saving the plot, you can edit it again through the SAS Explorer window or by selecting **File ▶ Open** from the ODS Graphics Editor window. Alternatively, you can reopen the saved plot for editing without first invoking SAS. For example, if you are running SAS under the Windows operating system, you can click on the plot to open it with the ODS Graphics Editor.

The ODS Graphics Editor does not permit you to make structural changes to a graph (such as moving the positions of data points). The ODS Graphics Editor provides you with a point-and-click way to make one-time changes to a specific graph, whereas the template language (see the section "Graph Templates" on page 687) provides you with a programmatic way to make template changes that persist every time you run the procedure. For complete details about the tools available in the ODS Graphics Editor, see *SAS/GRAPH: ODS Graphics Editor User's Guide*.

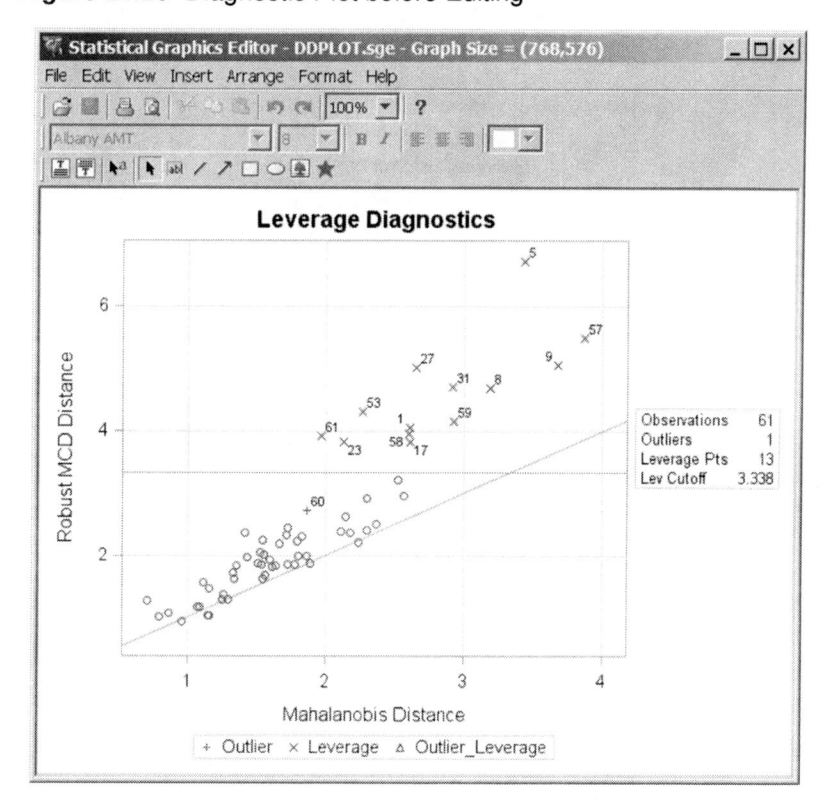

Figure 21.20 Diagnostic Plot before Editing

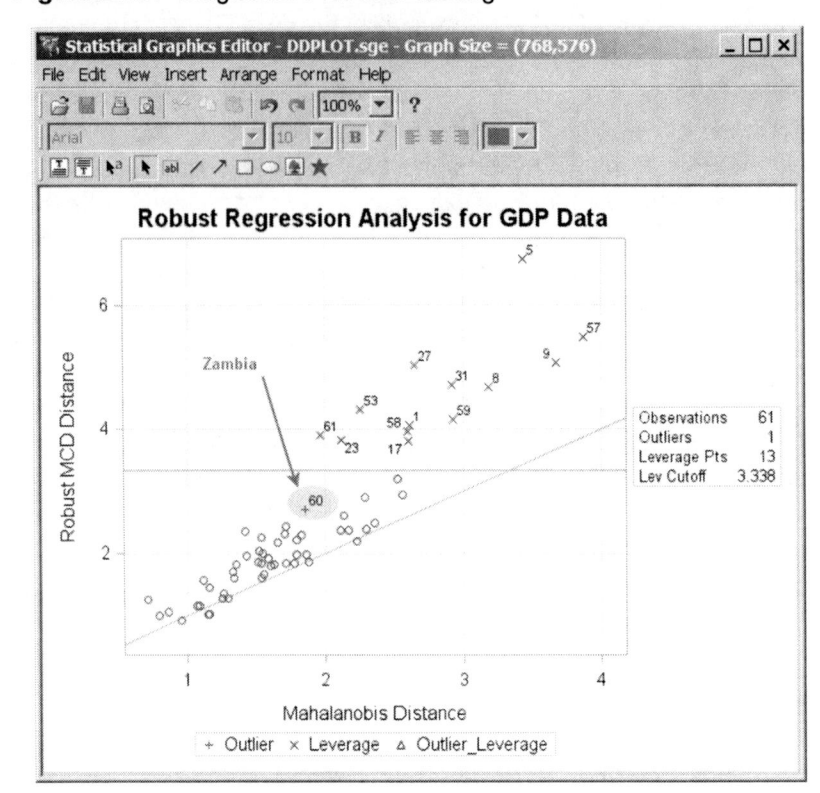

Figure 21.21 Diagnostic Plot after Editing

The Default Template Stores and the Template Search Path

Compiled templates are stored in a template store, which is a type of item store. (An item store is a special type of SAS file.) You can see the list of template stores by submitting the following statement:

```
ods path show;
```

The results are as follows:

```
Current ODS PATH list is:

1. SASUSER.TEMPLAT(UPDATE)
2. SASHELP.TMPLMST(READ)
```

These results show that the default template search path consists of SASUSER.Templat followed by SASHELP.Tmplmst. You can add template stores that you create or change the order in which the template stores are searched. This is discussed in detail in the sections "Saving Customized Templates" on page 697, "Using Customized Templates" on page 697, and "Reverting to the Default Templates" on page 698.

This section discusses the default template stores that you use when you have not modified the template search path with the ODS PATH statement. By default, templates that you write are stored in SASUSER.Templat. If you stored a modified template in SASUSER.Templat, ODS finds and uses your modified template. Otherwise, ODS finds the templates it provides in SASHELP.Tmplmst. You can see a list of all of the templates that you have modified as follows:

```
proc template;
   list / store=sasuser.templat;
run;
```

You can delete any template that you modified (so that ODS finds the default SAS template) by specifying it in a DELETE statement, as in the following statement:

```
proc template;
   delete Stat.REG.Graphics.ResidualPlot;
run;
```

ODS never deletes a template in SASHELP.Tmplmst, so you can safely run the preceding step, even if the template you specify does not exist in SASUSER.Templat. You can run the following step to delete the entire SASUSER.Templat template store of customized templates so that ODS uses only the SAS supplied templates:

```
ods path sashelp.tmplmst(read);
proc datasets library=sasuser nolist;
   delete templat(memtype=itemstor);
run;
ods path sasuser.templat(update) sashelp.tmplmst(read);
```

It is good practice to delete templates that you have customized when you are done with them, so that they are not unexpectedly used later. See the section "Reverting to the Default Templates" on page 698 for more information.

Styles

ODS styles control the overall appearance of your output. Usually, the only thing you need to do with styles is specify them in an ODS destination statement, as in the following example:

```
ods html body='b.html' style=Statistical;
```

However, you can also modify existing styles and even write your own styles. You can also specify style elements in custom templates that you write, or you can modify which style elements are used in templates supplied by SAS. This section provides an overview of styles and style elements, which are the components of a style. It also describes how to customize a style definition and how to specify a default style for your output. Only the most commonly used styles, style elements, and style changes are discussed here. For complete details about styles, see the *SAS Output Delivery System: User's Guide*.

An Overview of Styles

An ODS style definition provides formatting information for specific visual aspects of your SAS output (see the section "Style Elements and Attributes" on page 666). The appearance of tables and graphs is coordinated within a particular style. For tables, this information typically includes a list of font definitions and a list of colors. Each font definition specifies a family, size, weight, and style. Colors are associated with common areas of output, including titles, footnotes, BY groups, table headers, and table cells. For graphs, styles also control the appearance of graph elements including lines, markers, fonts and colors. ODS styles also include elements specific to statistical graphics, such as the style of fitted lines, confidence bands, and prediction limits. For more information about styles, see the *SAS Output Delivery System: User's Guide*.

You can specify a style by using the STYLE= option in an ODS destination statement such as HTML, PDF, RTF, or PRINTER. You can also specify a style in the LISTING destination; however, it affects graphs but not tables. Each style produces output with the same content, but with a different visual appearance. For example, the following statement requests output produced with the JOURNAL style:

```
ods html style=Journal;
```

You can use any SAS style or any style that you define yourself. The following statements list the names of all of the styles and then display four of them:

```
proc template;
   list styles;
   source Styles.Default;
   source Styles.Statistical;
   source Styles.Journal;
   source Styles.RTF;
run;
```

The results of this step (not shown) are a list of over fifty styles in the SAS listing and four style definitions in the SAS log. The style definitions are often hundreds of lines long. See the section "Style Definitions and Colors" on page 667 for more information about style definitions. While you can use any style, only seven styles are typically used with ODS Graphics. They are described in Table 21.2.

Table 21.2 Styles

Style	Default in	Description
DEFAULT	HTML	A color style whose dominant colors are blue, gray, and white, with bold sans-serif fonts. See Figure 21.13.
STATISTICAL	SAS/STAT documentation	A color style whose dominant colors are blue, creamy gray, and white, with sans-serif fonts. See Figure 21.14.
LISTING	LISTING	A color style, similar to DEFAULT, but with a white background. See Figure 21.12.
JOURNAL		A black-and-white style with filled areas, and with sans-serif fonts. See Output 21.3.1.
JOURNAL2		A black-and-white style, similar to JOURNAL, but with empty areas. See Output 21.3.2.
RTF	RTF	A color style whose dominant colors are blue, white, and black, with Times Roman fonts. See Figure 21.11.
ANALYSIS		A color style, similar to STATISTICAL, whose dominant color is tan. See Figure 21.8.

Each ODS destination has its own default style, as shown in Table 21.2. Most output in SAS/STAT documentation uses the STATISTICAL style. However, throughout this chapter, you can see examples of other styles. For more information about styles, see the *SAS Output Delivery System: User's Guide*.

Style Elements and Attributes

An ODS style definition is composed of a set of *style elements*. A style element is a collection of *style attributes* that applies to a particular feature or aspect of the output. A value is specified for each attribute in a style definition. For example, `GraphFit` is the style element used for fit lines, and its attributes include: `LineThickness`, `LineStyle`, `MarkerSize`, `MarkerSymbol`, `ContrastColor`, and `Color`.

In general, style definitions control the overall appearance of ODS tables and graphs. For tables, style definitions specify features such as background color, table borders, and color scheme; and they specify the fonts, sizes, and color for the text and values in a table and its headers. For graphs, style definitions specify the following features:

- background color
- graph dimensions (height and width)
- borders
- line styles for axes and grid lines
- fonts, sizes, and colors for titles, footnotes, axis labels, axis values, and data labels. See the section "Modifying Graph Fonts in Styles" on page 680 for an illustration.
- marker symbols, colors, and sizes for data points and outliers
- line styles for needles
- line and curve styles for fitted models and predicted values. See "Modifying Other Graph Elements in Styles" on page 683 for an illustration.
- line and curve styles for confidence and prediction limits
- fill colors for histogram bars, confidence bands, and confidence ellipses
- colors for box plot features
- colors for surfaces
- color ramps for contour plots

SAS supplies a template for each plot that is created by statistical procedures. A graph template is a program that specifies the layout and details of a graph. See the section "Graph Templates" on page 687 for more information about templates. Some template options are specified with a style reference of the form `style-element`, or occasionally `style-element:attribute`. For example, the symbol, color, and size of markers for basic scatter plots are specified in a template SCATTERPLOT statement as follows:

```
scatterplot x=x y=y / markerattrs=GraphDataDefault;
```

The preceding statement specifies that the appearance for markers is controlled by the `GraphDataDefault` element. Consistent use of this element guarantees a common appearance of markers across all scatter plots, based on the style definition that you are using.

In general, ODS Graphics features are determined by style element attributes unless they are overridden by a statement or option in the graph template. For example, suppose that a classification variable is specified with the GROUP= option in a SCATTERPLOT template statement as follows:

```
scatterplot x=X y=Y / group=GroupVar;
```

Then the colors for markers that correspond to the classification levels are assigned by using the style element attributes `GraphData1:ContrastColor` through `GraphData12:ContrastColor`.

Style definitions are created and modified with PROC TEMPLATE. For more information, see the *SAS Output Delivery System: User's Guide.* You need to understand the relationships between style elements and graph features if you want to create your own style definition or modify a style definition. This is explained in the following sections.

Style Definitions and Colors

The default style definitions that the SAS System provides are stored in the *Styles* directory of SASHELP.Tmplmst. You can display, edit, and save style definitions by using the same methods available for modifying template definitions, as explained in the section "The Default Template Stores and the Template Search Path" on page 662 and the series of sections beginning with "Displaying Templates" on page 693. In particular, you can display a style definition by using one of these methods:

- From the Templates window in the SAS windowing environment, expand the SASHELP.Tmplmst node under **Templates**, and then select **Styles** to display the contents of this folder. To open the Templates window, type **odst** in the command line.
- Use the SOURCE statement in PROC TEMPLATE.

For example, the following statements display the DEFAULT style definition in the SAS log:

```
proc template;
    source Styles.Default;
run;
```

Some of the results are as follows:

```
define style Styles.Default;
   . . .
   class GraphColors
      "Abstract colors used in graph styles" /
      . . .
      'gconramp3cend' = cxFF0000
      'gconramp3cneutral' = cxFF00FF
      'gconramp3cstart' = cx0000FF
      . . .
      'gdata12' = cxDDD17E
      'gdata11' = cxB7AEF1
      'gdata10' = cx87C873
      'gdata9' = cxCF974B
      'gdata8' = cxCD7BA1
      'gdata6' = cxBABC5C
      'gdata7' = cx94BDE1
      'gdata4' = cxA9865B
      'gdata5' = cxB689CD
      'gdata3' = cx66A5A0
      'gdata2' = cxDE7E6F
      'gdata1' = cx7C95CA;
   . . .
```

The first part of this list shows that the shading for certain filled plots, such as some contour plots goes, from blue (`'gconramp3cstart' = cx0000FF`) to magenta (`'gconramp3cneutral' = cxFF00FF`) to red (`'gconramp3cend' = cxFF0000`). All colors are specified in values of the form CX*rrggbb*, where the last six characters specify RGB (red, green, blue) values on the hexadecimal scale of 00 to FF (or 0 to 255 base 10). The second part of the list (`'gdata1' = cx7C95CA`) shows that the dominant component of the `GraphData1` color is blue because the blue component of the color (CA, which corresponds to 202 base 10) is greater than both the green component (95, which corresponds to 149 base 10) and the red component (7C, which corresponds to 124 base 10).

You can change any part of the style and then submit the style back into SAS, after first submitting a PROC TEMPLATE statement. See the sections "Saving Customized Templates" on page 697, "Using Customized Templates" on page 697, and "Reverting to the Default Templates" on page 698 for more information about modifying, using, and restoring templates. The principles discussed in those sections apply to all templates—table, style, and graph.

Some Common Style Elements

The DEFAULT style is the parent for the styles used for statistical graphics work. You can see all of the elements of the DEFAULT style by running the following step:

```
proc template;
   source styles.default;
run;
```

The source listing of the definition of the DEFAULT style is hundreds of lines long. If you run PROC TEMPLATE with the SOURCE statement for most other styles, you see `parent = styles.default`, and you do not see all of the elements in the style unless you also run the preceding step.

Only a few of the style elements are referenced in the templates that the SAS System provides for statistical procedures. The most commonly used style elements, along with the defaults for the noncolor attributes of the DEFAULT style, are shown next (`Color` applies to filled areas, and `ContrastColor` applies to markers and lines):

`Graph`	graph size, outer border appearance, and background color
	`Padding = 0`
	`BackgroundColor`
`GraphConfidence`	graph size, outer border appearance, and background color
	`LineThickness = 1px`
	`LineStyle = 1`
	`MarkerSize = 7px`
	`MarkerSymbol = "triangle"`
	`ContrastColor`
	`Color`
`GraphData1`	attributes related to first grouped data items
	`MarkerSymbol = "circle"`
	`LineStyle = 1`
	`ContrastColor`
	`Color`
`GraphData2`	attributes related to second grouped data items
	`MarkerSymbol = "plus"`
	`LineStyle = 4`
	`ContrastColor`
	`Color`
`GraphData3`	attributes related to third grouped data items
	`MarkerSymbol = "X"`
	`LineStyle = 8`
	`ContrastColor`
	`Color`
`GraphData4`	attributes related to fourth grouped data items
	`MarkerSymbol = "triangle"`
	`LineStyle = 5`
	`ContrastColor`
	`Color`
`GraphData`*n*	attributes related to *n*th grouped data items
	`MarkerSymbol`
	`LineStyle`
	`ContrastColor`
	`Color`

`GraphDataDefault`	attributes related to non-grouped data items
	`EndColor`
	`NeutralColor`
	`StartColor`
	`MarkerSize = 7px`
	`MarkerSymbol = "circle"`
	`LineThickness = 1px`
	`LineStyle = 1`
	`ContrastColor`
	`Color`
`GraphFit`	primary fit line, such as a normal density curve
	`LineThickness = 2px`
	`LineStyle = 1`
	`MarkerSize = 7px`
	`MarkerSymbol = "circle"`
	`ContrastColor`
	`Color`
`GraphFit2`	secondary fit line, such as a kernel density curve
	`LineThickness = 2px`
	`LineStyle = 4`
	`MarkerSize = 7px`
	`MarkerSymbol = "X"`
	`ContrastColor`
	`Color`
`GraphGridLines`	horizontal and vertical grid lines drawn at major tick marks
	`Displayopts = "auto"`
	`LineThickness = 1px`
	`LineStyle = 1`
	`ContrastColor`
	`Color`
`GraphOutlier`	outlier data for the graph
	`LineThickness = 2px`
	`LineStyle = 42`
	`MarkerSize = 7px`
	`MarkerSymbol = "circle"`
	`ContrastColor`
	`Color`
`GraphPredictionLimits`	fills for prediction limits
	`LineThickness = 1px`
	`LineStyle = 2`
	`MarkerSize = 7px`
	`MarkerSymbol = "chain"`
	`ContrastColor`
	`Color`

`GraphReference`	horizontal and vertical reference lines and drop lines
	`LineThickness = 1px`
	`LineStyle = 1`
	`ContrastColor`
`GraphDataText`	text font and color for point and line labels
	`Font = GraphFonts('GraphDataFont')`
	(where `'GraphDataFont'` =
	`("<sans-serif>, <MTsans-serif>",7pt))`
	`Color`
`GraphValueText`	text font and color for axis tick values and legend values
	`Font = GraphFonts('GraphValueFont')`
	(where `'GraphValueFont'` =
	`("<sans-serif>, <MTsans-serif>",9pt))`
	`Color`
`GraphLabelText`	text font and color for axis labels and legend title
	`Font = GraphFonts('GraphLabelFont')`
	(where `'GraphLabelFont'` =
	`("<sans-serif>, <MTsans-serif>",10pt,bold))`
	`Color`
`GraphFootnoteText`	text font and color for footnote(s)
	`Font = GraphFonts('GraphFootnoteFont')`
	(where `'GraphFootnoteFont'` =
	`("<sans-serif>, <MTsans-serif>",10pt))`
	`Color`
`GraphTitleText`	text font and color for title(s)
	`Font = GraphFonts('GraphTitleFont')`
	(where `'GraphTitleFont'` = `("<sans-serif>,`
	`<MTsans-serif>",11pt,bold))`
	`Color`
`GraphWalls`	vertical wall(s) bounded by axes
	`LineThickness = 1px`
	`LineStyle = 1`
	`FrameBorder = on`
	`ContrastColor`
	`BackgroundColor`
	`Color`

You refer to these elements in graph templates as **style-element** or as **style-element:attribute** (for example `GraphDataDefault:ContrastColor`). The default values are not shown for the color attributes since they are typically defined indirectly. For example, `Graph:BackgroundColor` (the color that fills the box outside the graph) is defined elsewhere in the style as `colors('docbg')`. The style also defines: `'docbg' = color_list('bgA')` and `'bgA' = cxE0E0E0`. This shows that the background is a shade of gray that is much closer to white (CXFFFFFF) than to black (CX000000). You can see the background color in Figure 21.22. This shade of gray might seem darker (closer to CX000000) than you might expect based on just the RGB values. Your perception of a color change is not a linear function of the change in RGB values.

You can use the following program to see the color and other attributes for a number of style elements:

```
proc format; value vf 5 = 'GraphValueText'; run;

data x;
   array y[20] y0 - y19;
   do x = 1 to 20; y[x] = x - 0.5; end;
   do x = 0 to 10 by 5; output; end;
   label y0 = 'GraphLabelText' x = 'GraphLabelText';
   format x y0 vf.;
run;

%macro d;
   %do i = 1 %to 12;
      reg y=y%eval(19-&i) x=x / lineattrs=GraphData&i markerattrs=GraphData&i
                                curvelabel="  GraphData&i" curvelabelpos=max;
   %end;
%mend;

%macro l(i, l);
   reg y=y&i x=x / lineattrs=&l markerattrs=&l curvelabel="  &l"
                   curvelabelpos=max;
%mend;

ods listing style=default;

proc sgplot noautolegend;
   title 'GraphTitleText';
   %d
   %l(19, GraphDataDefault)
   %l( 6, GraphFit)
   %l( 5, GraphFit2)
   %l( 4, GraphPredictionLimits)
   %l( 3, GraphConfidence)
   %l( 2, GraphGridLines)
   %l( 1, GraphOutlier)
   %l( 0, GraphReference)
   xaxis values=(0 5 10);
run;
```

The results in Figure 21.22 display the attributes for a number of the elements of the DEFAULT style.

Figure 21.22 Attributes of Style Elements in the DEFAULT Style

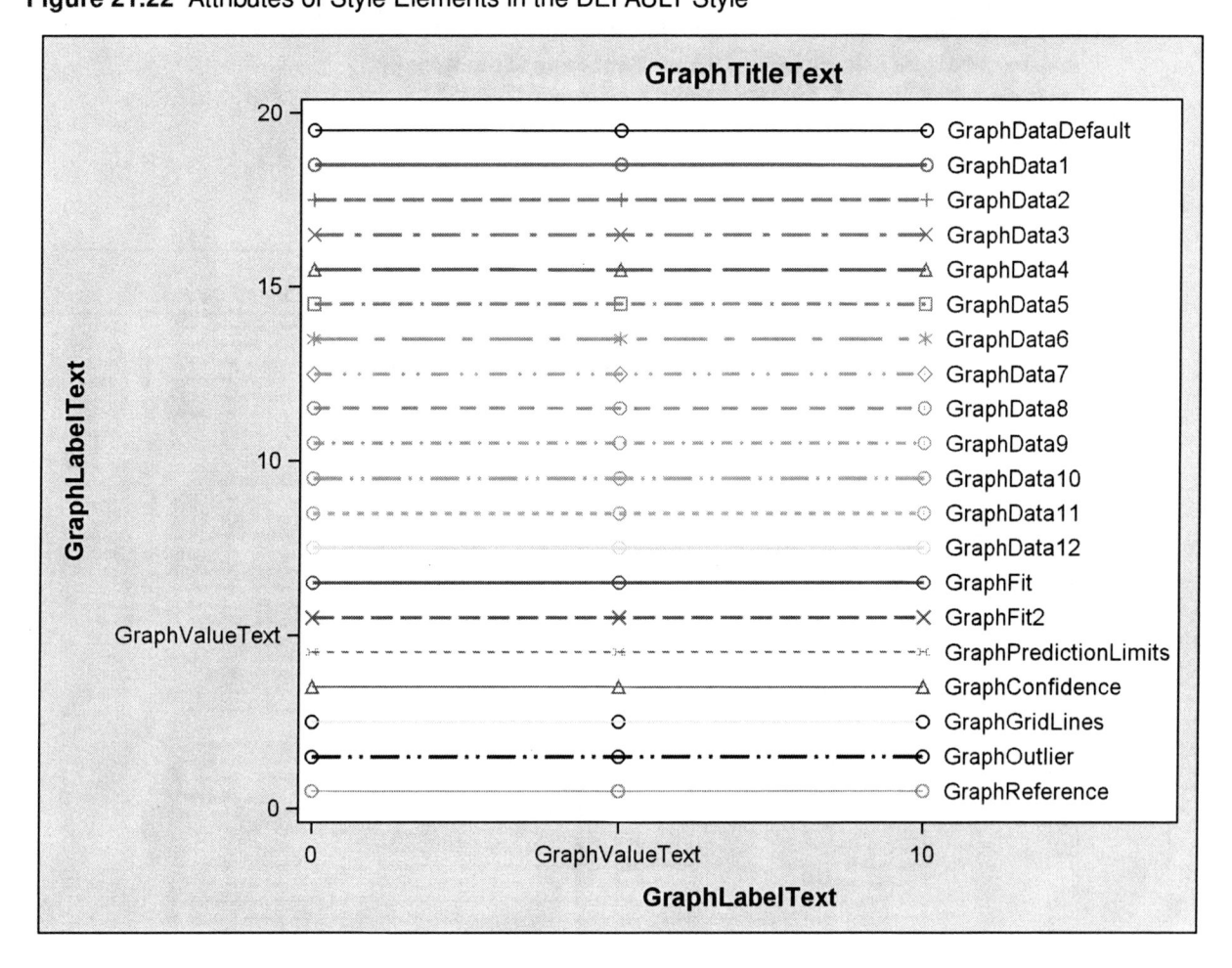

When there is a group or classification variable, the colors, markers, and lines that distinguish the groups are derived from the `GraphData`*n* elements that are defined with the style. In the DEFAULT style, these are elements `GraphData1` through `GraphData12`. There can be any number of groups even though only 12 `GraphData`*n* style elements are defined in the DEFAULT style. The following steps create a data set with 40 groups, display one line per group, and produce Figure 21.23:

```
data x;
   do y = 40 to 1 by -1;
      group = 'Group' || put(41 - y, 2. -L);
      do x = 0 to 10 by 5;
         if x = 10 then do; z = 11; l = group; end;
         else            do; z = .;  l = ' ';   end;
         output;
      end;
   end;
run;
```

Chapter 21: Statistical Graphics Using ODS

```
proc sgplot data=x;
   title 'Colors, Markers, Lines Patterns for Groups';
   series  y=y x=x / group=group markers;
   scatter y=y x=z / group=group markerchar=l;
run;
```

Figure 21.23 Markers and Lines Cycle with Different Periods in Groups

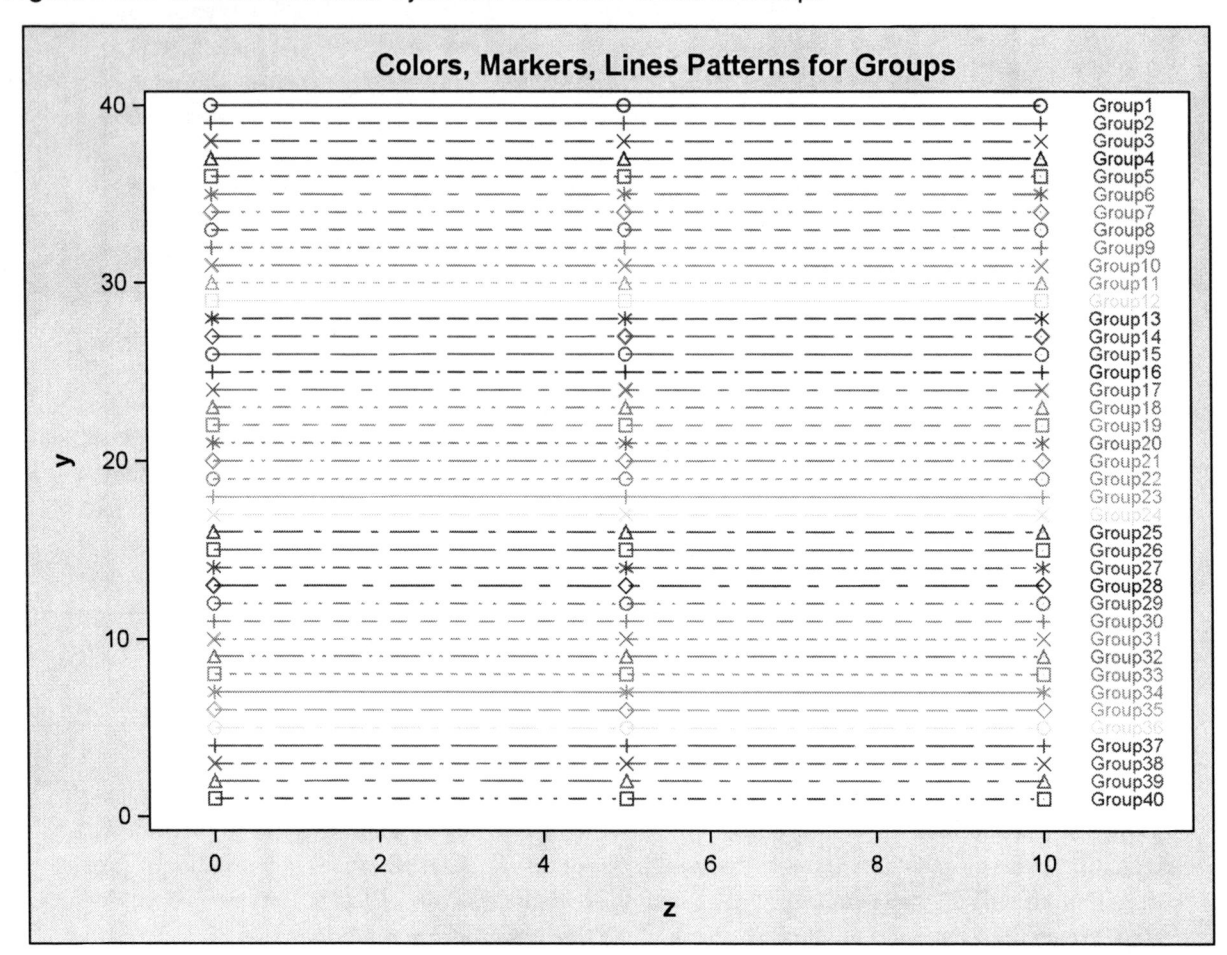

The colors, markers, and line patterns in Figure 21.23 repeat in cycles. The `GraphData1` – `GraphData8` lines in Figure 21.22 exactly match the `Group1` – `Group8` lines in Figure 21.23. After that, there are differences due to the cyclic construction of the grouped style definition. This is explained next.

The DEFAULT style defines a marker symbol only in `GraphData1` through `GraphData7`. The seven markers are: circle, plus, X, triangle, square, asterisk, and diamond. With the explicit style reference in Figure 21.22, the actual symbol, when no symbol is specified, is the circle. This is what you see for `GraphData8` through `GraphData12`. With the group variable in Figure 21.23, the symbols repeat in cycles. Hence, `Group1`, `Group8`, `Group15`, and so on, are all circles. Similarly, `Group2`, `Group9`, `Group16`, and so on, are all pluses. The DEFAULT style defines 11 different line styles for `GraphData1` through `GraphData11`: 1, 4, 8, 5, 14, 26, 15, 20, 41, 42, and 2. Hence, `Group1`, `Group12`, `Group23`, and so on, all have the same line style, which is a solid line. Similarly, `Group2`, `Group13`, `Group24`, and so on, all have line style 4. There are twelve different colors, so `Group1`, `Group13`, `Group25`, and so on, all have the same colors. Overall, there are $12 \times 11 \times 7 = 924$ color/line/marker combinations that appear before any combination repeats. You can use the %MODSTYLE SAS autocall macro (see the sections "Creating an All-Color Style by Using the ModStyle Macro" on page 675 and "Style Template Modification Macro" on page 699) to conveniently change these style attributes.

Creating an All-Color Style by Using the ModStyle Macro

Many styles are designed to make color plots where lines, functions, and groups of observations can be distinguished even when the plot is sent to a black-and-white printer. Hence, lines differ not only in color but also in pattern. Similarly, markers differ in color and symbol.

You can change that behavior with the %MODSTYLE autocall macro. It creates a new style by modifying a parent style and reordering the colors, line patterns, and marker symbols in the `GraphData`*n* style elements (see section "Some Common Style Elements" on page 668). By default, the macro creates a new style that distinguishes lines and groups only by color. The macro is documented in section "Style Template Modification Macro" on page 699.

The following example illustrates the default use of the macro and is taken from the section "Fitting a Curve through a Scatter Plot" on page 7601 of Chapter 91, "The TRANSREG Procedure." The data come from an experiment in which nitrogen oxide emissions from a single cylinder engine are measured for various combinations of fuel and equivalence ratio. The following statements create the SAS data set:

```
data Gas;
   input Fuel :$8. EqRatio NOx @@;
   label EqRatio = 'Equivalence Ratio'
         NOx     = 'Nitrogen Oxide Emissions';
   datalines;
Ethanol  0.907 3.741 Ethanol  0.761 2.295 Ethanol  1.108 1.498

   ... more lines ...

;
```

The following statements fit separate curves for each group and produce Figure 21.24 and Figure 21.25:

```
ods listing style=statistical;
ods graphics on;

proc transreg data=Gas ss2 plots=transformation lprefix=0;
   model identity(nox) = class(Fuel / zero=none) * pbspline(EqRatio);
run;

%modstyle(parent=statistical, name=StatColor)
ods listing style=StatColor;

proc transreg data=Gas ss2 plots=transformation lprefix=0;
   model identity(nox) = class(Fuel / zero=none) * pbspline(EqRatio);
run;
```

The first PROC TRANSREG step uses the STATISTICAL style to create the fit plot in Figure 21.24, which uses different colors, line patterns, and markers for each group. Then the macro creates a new style, called STATCOLOR, that inherits its characteristics from the STATISTICAL style. Only the attributes of the lines and markers are changed. In Figure 21.25, which is created with the modified style, the groups are differentiated only by color. This is the easiest and most common way for you to use this macro. However, you can use it to perform other style modifications as illustrated in the section "Changing the Default Markers and Lines" on page 677. The macro is documented in section "Style Template Modification Macro" on page 699.

Figure 21.24 Fit Plot with the STATISTICAL Style

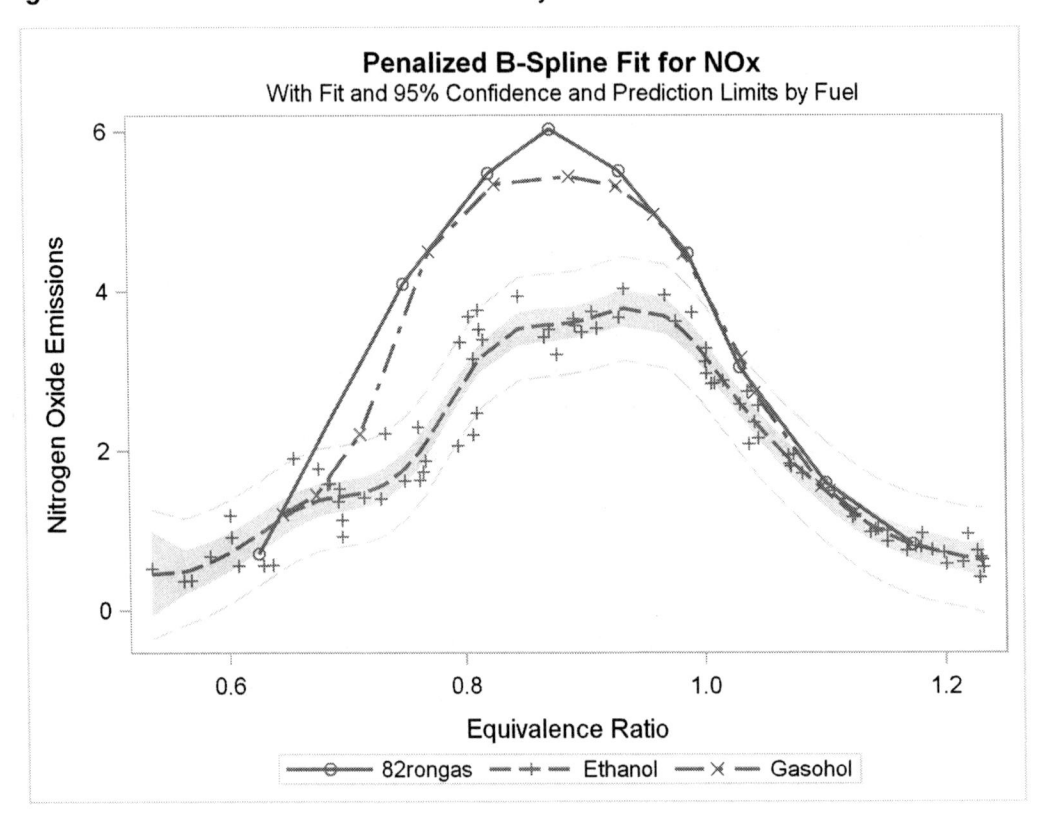

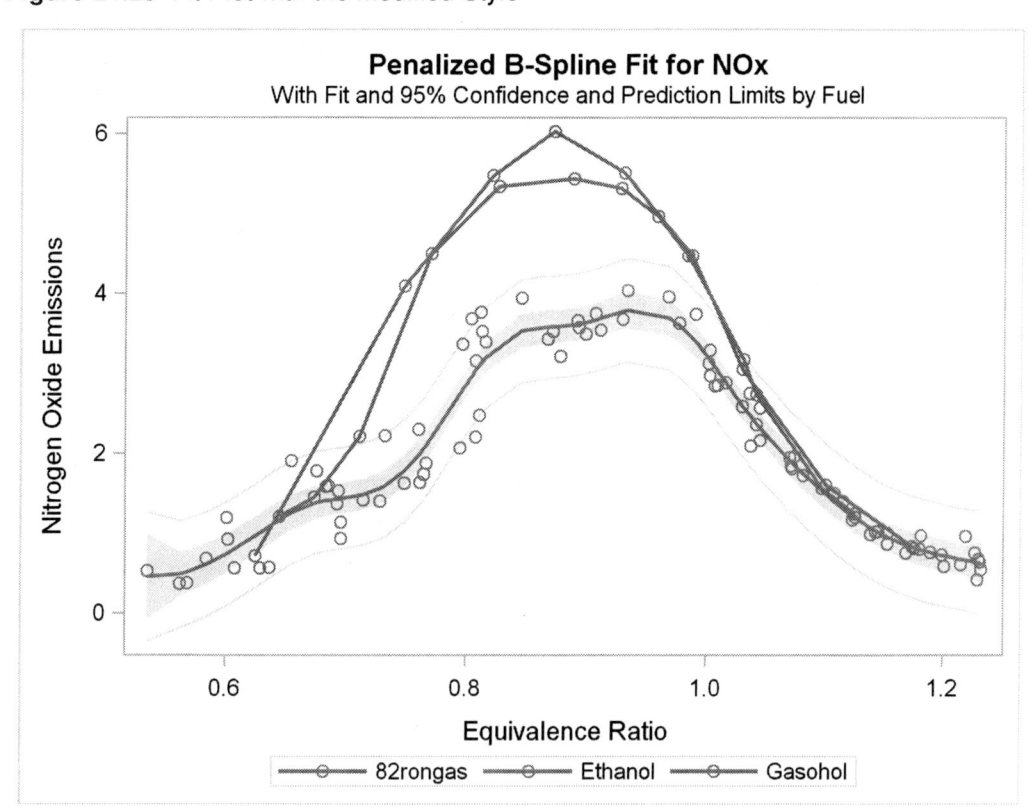

Figure 21.25 Fit Plot with the Modified Style

Penalized B-Spline Fit for NOx
With Fit and 95% Confidence and Prediction Limits by Fuel

Changing the Default Markers and Lines

The preceding section shows how to use the %MODSTYLE autocall macro to create an all-color style. You can also use the %MODSTYLE macro to change markers and line styles. This example creates a new style called MARKSTYLE that inherits from the STATISTICAL style but uses a different set of markers. The following statements create artificial data, change the marker list, and display the results:

```
data x;
   do g = 1 to 12;
      do x = 1 to 10;
         y = 13 - g + sin(x * 0.1 * g);
         output;
      end;
   end;
run;

%modstyle(name=markstyle, parent=statistical, type=CLM,
          markers=star plus circle square diamond starfilled
                  circlefilled squarefilled diamondfilled)

ods listing style=markstyle;

proc sgplot;
   title 'Modified Marker List';
   loess y=y x=x / group=g;
run;
```

The NAME= option specifies the new style name, and the PARENT= option specifies the parent style. The TYPE= option controls the method of cycling through colors, lines, and markers. The default, TYPE=LMbyC, fixes (holds constant) the line styles and markers, while cycling through the color list. This is illustrated in the section "Creating an All-Color Style by Using the ModStyle Macro" on page 675. This example uses TYPE=CLM to cycle through colors, line styles, and markers (holding none of them fixed). Other TYPE= values are described in section "Style Template Modification Macro" on page 699. The values specified with the TYPE= option are case sensitive ('by' is lower case and the 'L', 'C', and 'M' are upper case). The new marker list is specified with the MARKERS= option. The results are displayed in Figure 21.26. The marker list is reused in the tenth and subsequent groups since only nine markers are defined.

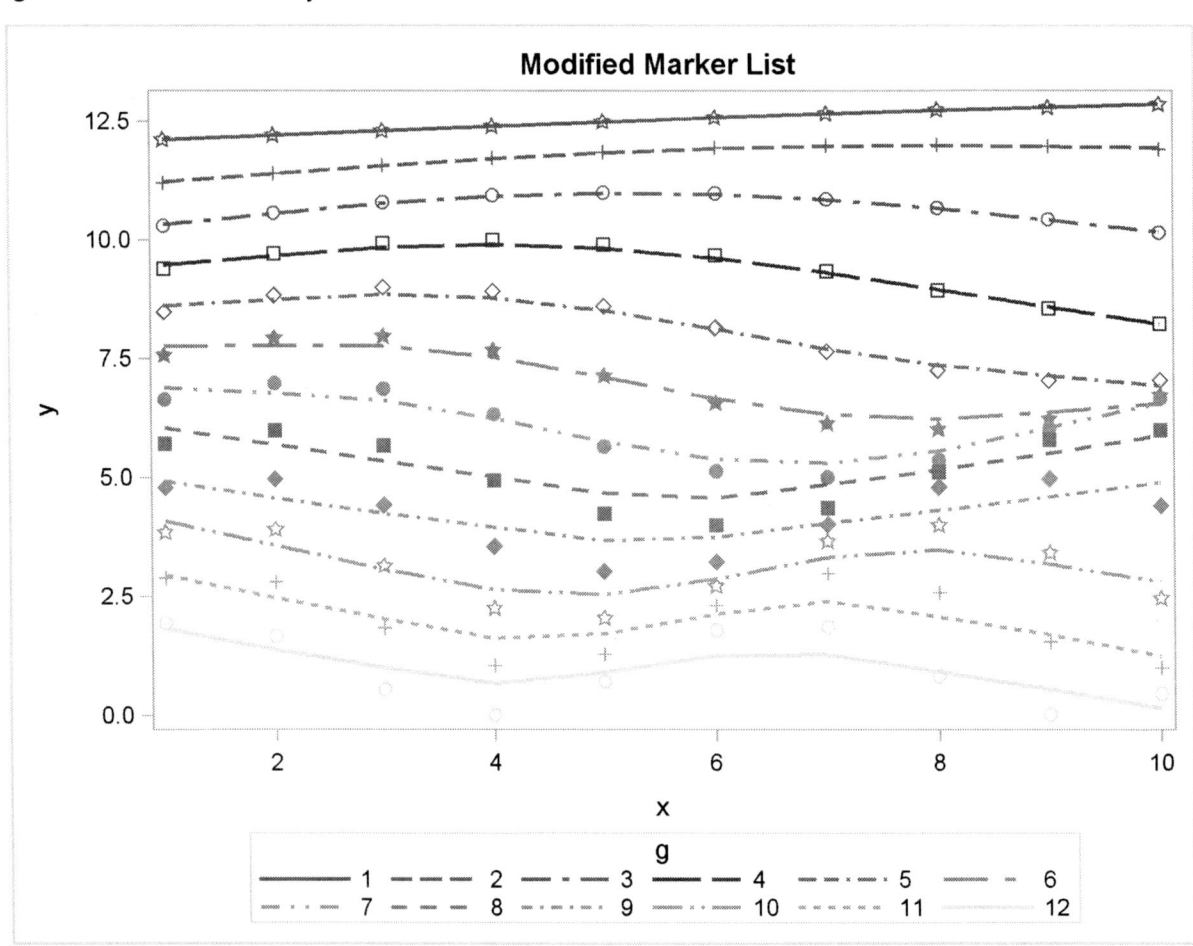

Figure 21.26 A Modified Style with a New List of Markers

The following statements create a new style called LINESTYLE that inherits from the STATISTICAL style and changes the line list:

```
%modstyle(name=linestyle, parent=statistical, type=CLM,
          linestyles=Solid LongDash MediumDash Dash ShortDash Dot ThinDot)

ods listing style=linestyle;
```

```
proc sgplot;
   title 'Modified Line Style List';
   loess y=y x=x / group=g;
run;
```

The new line list is specified with the LINESTYLES= option. The results are displayed in Figure 21.27. In this example, each of the first seven groups uses a dash pattern that is shorter than the previous group. The line list is reused in the eighth and subsequent groups since only seven line patterns are defined.

Figure 21.27 Modified Style with a New List of Line Styles

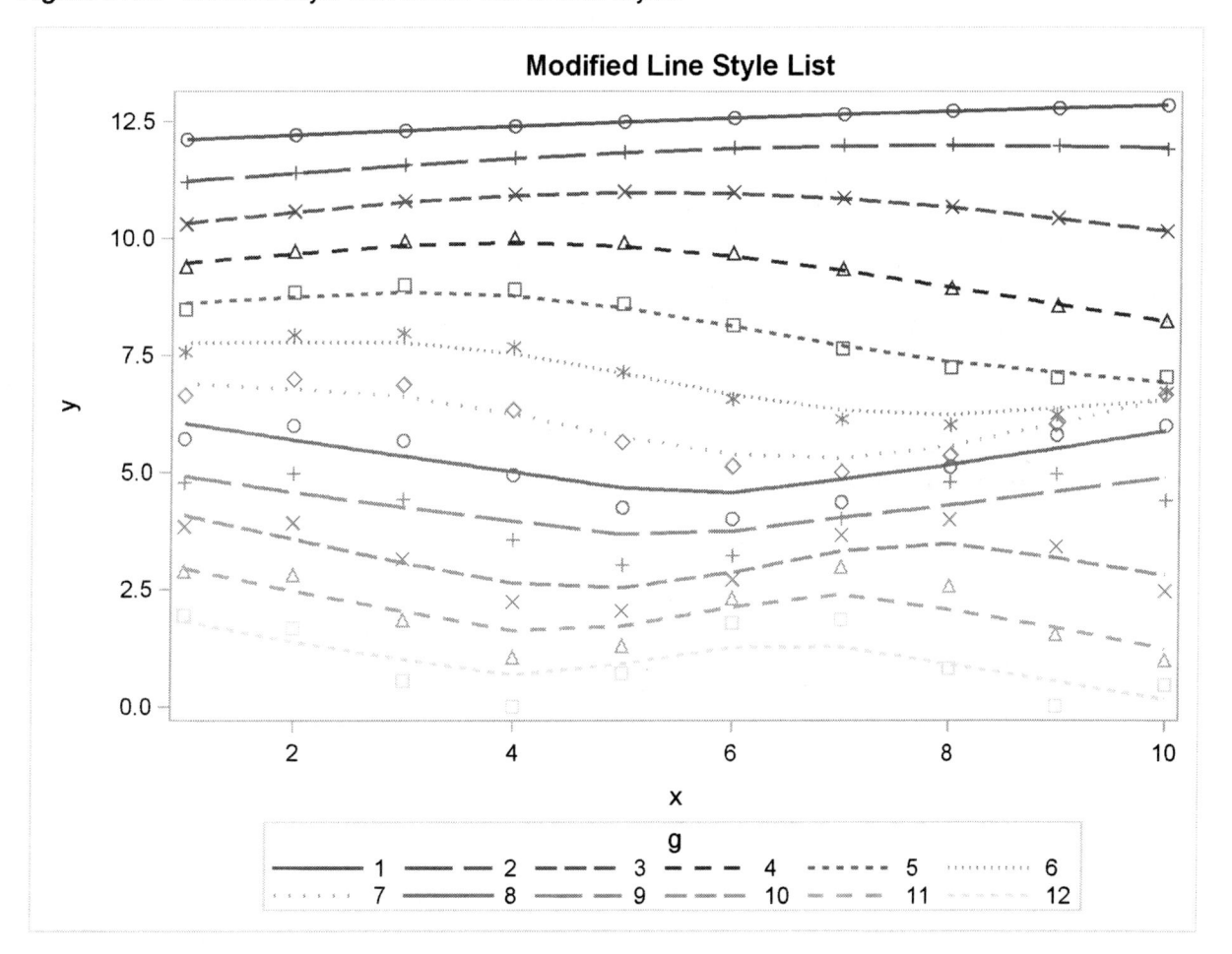

You can learn more about style modification by examining the new styles, as in the following example:

```
proc template;
   source styles.markstyle;
   source styles.linestyle;
run;
```

The results show the definitions of `GraphData1` through `GraphData32` that the macro created. An abridged listing of the results follows:

```
define style Styles.Markstyle;
   parent = Styles.statistical;
   . . .
   style GraphData1 /
      markersymbol = "star"
      linestyle = 1
      contrastcolor = ColorStyles('c1')
      color = FillStyles('f1');
   . . .
   style GraphData32 /
      markersymbol = "diamond"
      linestyle = 42
      contrastcolor = ColorStyles('c8')
      color = FillStyles('f8');
end;

define style Styles.Linestyle;
   parent = Styles.statistical;
   . . .
   style GraphData1 /
      markersymbol = "circle"
      linestyle = 1
      contrastcolor = ColorStyles('c1')
      color = FillStyles('f1');
   . . .
   style GraphData32 /
      markersymbol = "triangle"
      linestyle = 20
      contrastcolor = ColorStyles('c8')
      color = FillStyles('f8');
end;
```

You can use the NUMBEROFGROUPS= option in the %MODSTYLE macro to control the number of `GraphData`*n* style elements created in the new style.

Modifying Graph Fonts in Styles

You can modify an ODS style to customize the general appearance of plots produced with ODS Graphics, just as you can modify a style to customize the general appearance of ODS tables. This section shows you how to customize fonts used in graphs. The following statements show the STATISTICAL style:

```
proc template;
   source Styles.Statistical;
run;
```

The portion of the style that controls fonts is shown next:

```
style GraphFonts /
   'GraphDataFont' = ("<sans-serif>, <MTsans-serif>",7pt)
   'GraphUnicodeFont' = ("<MTsans-serif-unicode>",9pt)
   'GraphValueFont' = ("<sans-serif>, <MTsans-serif>",9pt)
```

```
    'GraphLabelFont'    = ("<sans-serif>, <MTsans-serif>",10pt)
    'GraphFootnoteFont' = ("<sans-serif>, <MTsans-serif>",10pt,italic)
    'GraphTitleFont'    = ("<sans-serif>, <MTsans-serif>",11pt,bold)
    'GraphAnnoFont'     = ("<sans-serif>, <MTsans-serif>",10pt);
```

The following fonts are the ones typically used for the text in most graphs:

- `GraphDataFont` is the smallest font. It is used for text that needs to be small (labels for points in scatter plots, labels for contours, and so on)

- `GraphValueFont` is the next largest font. It is used for axis value (tick marks) labels and legend entry labels.

- `GraphLabelFont` is the next largest font. It is used for axis labels and legend titles.

- `GraphFootnoteFont` is the next largest font. It is used for all footnotes.

- `GraphTitleFont` is the largest font. It is used for all titles.

The following statements define a style named NEWSTYLE that replaces the graph fonts in the DEFAULT style with italic Times New Roman fonts, which are available with the Windows operating system:

```
proc template;
   define style Styles.NewStyle;
      parent=Styles.Statistical;
      replace GraphFonts /
         'GraphDataFont'     = ("<MTserif>, Times New Roman",7pt)
         'GraphUnicodeFont'  = ("<MTserif>, Times New Roman",9pt)
         'GraphValueFont'    = ("<MTserif>, Times New Roman",9pt)
         'GraphLabelFont'    = ("<MTserif>, Times New Roman",10pt)
         'GraphFootnoteFont' = ("<MTserif>, Times New Roman",10pt)
         'GraphTitleFont'    = ("<MTserif>, Times New Roman",11pt)
         'GraphAnnoFont'     = ("<MTserif>, Times New Roman",10pt);
   end;
run;
```

For more information about the DEFINE, PARENT, and REPLACE statements, see the *SAS/GRAPH: Graph Template Language Reference*.

The "Getting Started" section of Chapter 75, "The ROBUSTREG Procedure," creates the following data set to illustrate the use of the PROC ROBUSTREG for robust regression:

```
data stack;
   input  x1 x2 x3 y @@;
   datalines;
80  27  89  42     80  27  88  37     75  25  90  37

   ... more lines ...

;
```

682 ✦ *Chapter 21: Statistical Graphics Using ODS*

The following statements create a Q-Q plot that uses the STATISTICAL style (see Figure 21.28) and the NEWSTYLE style (see Figure 21.29):

```
ods listing style=Statistical;
ods graphics on;

proc robustreg data=stack plots=qqplot;
   ods select QQPlot;
   model y = x1 x2 x3;
run;

ods listing close;
ods listing style=NewStyle;

proc robustreg data=stack plots=qqplot;
   ods select QQPlot;
   model y = x1 x2 x3;
run;
```

Figure 21.28 Q-Q Plot That Uses the STATISTICAL Style

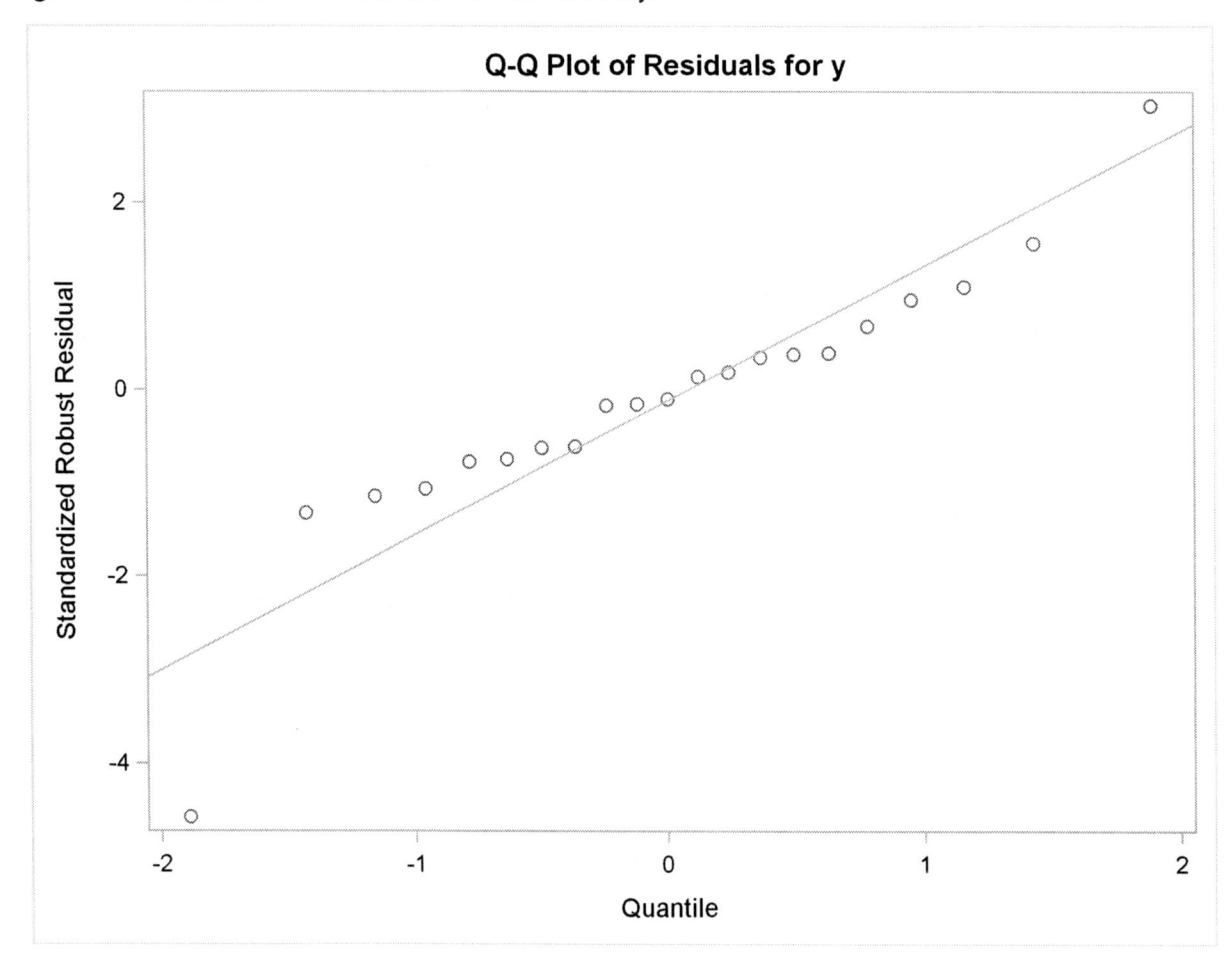

Figure 21.29 Q-Q Plot That Uses the NEWSTYLE Style

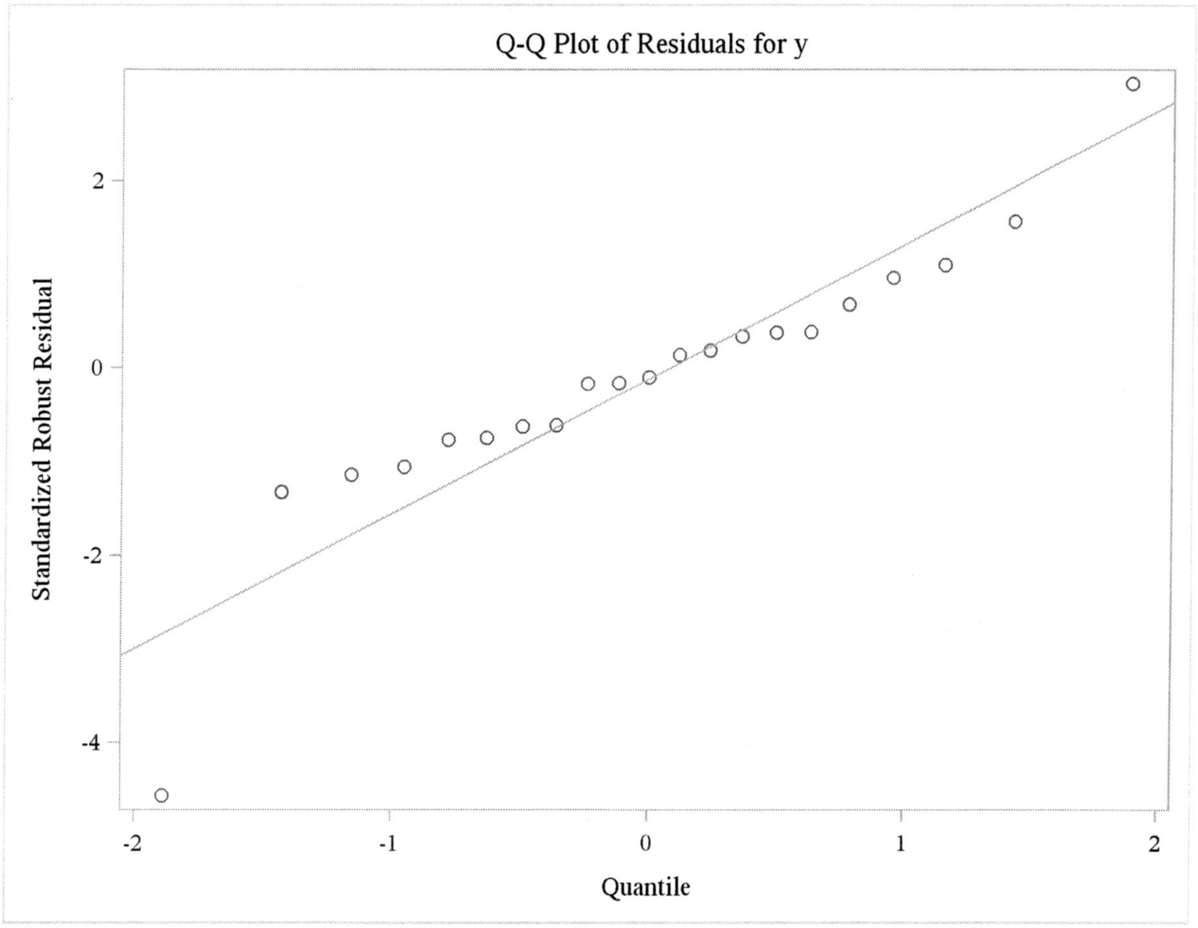

Although this example illustrates the use of a style with graphical output from a particular procedure, a style is applied to *all* of your output (graphs and tables) in the destination for which you specify the style. See the section "Changing the Default Style" on page 685 for information about specifying a default style for all your output.

Modifying Other Graph Elements in Styles

This section illustrates how to modify other style elements for graphics, specifically the style element `GraphReference`, which controls the attributes of reference lines. You can run the following statements to learn more about the `GraphReference` style element:

```
proc template;
   source styles.statistical;
run;
```

The following are the first two lines of the source listing:

```
define style Styles.Statistical;
   parent = styles.default;
```

There is no mention of `GraphReference` in the complete listing of the source because `GraphReference` is inherited from the parent style. Most styles inherit many of their attributes from other styles. To find out more, you must list the parent style, as in the following example:

```
proc template;
   source styles.default;
run;
```

Styles that you typically use with ODS Graphics inherit most of their attributes from only one style, the DEFAULT style. A few of the other styles inherit from several parents. You might have to repeat this process multiple times to find the first parent. However, with the STATISTICAL style, you only need to run this one extra step, and the results contain the following:

```
class GraphReference /
   linethickness = 1px
   linestyle = 1
   contrastcolor = GraphColors('greferencelines');
```

To specify a line thickness of 4 pixels for all reference lines, add the following statement to the definition of the NEWSTYLE style in the section "Modifying Graph Fonts in Styles" on page 680:

```
replace GraphReference / linethickness=4px;
```

The following statements modify the style and produce the Q-Q plot shown in Figure 21.30:

```
proc template;
   define style Styles.NewStyle;
      parent=Styles.Statistical;
      replace GraphFonts     /
         'GraphDataFont'     = ("<MTserif>, Times New Roman",7pt)
         'GraphUnicodeFont'  = ("<MTserif>, Times New Roman",9pt)
         'GraphValueFont'    = ("<MTserif>, Times New Roman",9pt)
         'GraphLabelFont'    = ("<MTserif>, Times New Roman",10pt)
         'GraphFootnoteFont' = ("<MTserif>, Times New Roman",10pt)
         'GraphTitleFont'    = ("<MTserif>, Times New Roman",11pt)
         'GraphAnnoFont'     = ("<MTserif>, Times New Roman",10pt);
      replace GraphReference / linethickness=4px;
   end;
run;

ods listing style=NewStyle;
ods graphics on;

proc robustreg data=stack plots=qqplot;
   ods select QQPlot;
   model y = x1 x2 x3;
run;
```

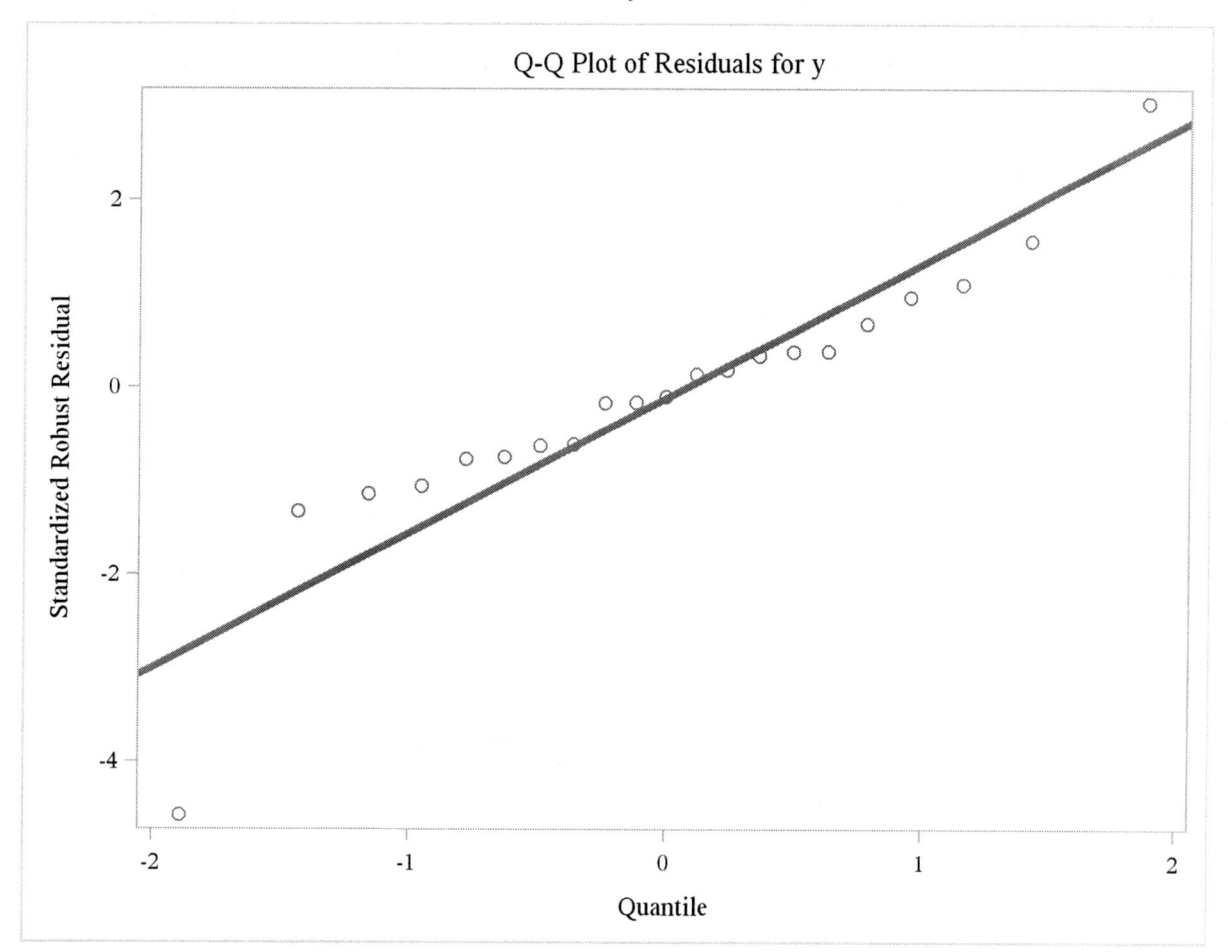

Figure 21.30 Q-Q Plot That Uses the NEWSTYLE Style with a Thicker Line

You can use this approach to modify other attributes of the line, such as `LineStyle` and `ContrastColor`. These style modifications apply to all graphs that display reference lines, and not just to Q-Q plots produced by PROC ROBUSTREG. You can control the attributes of specific graphs by modifying the graph template, as discussed in the section "Graph Templates" on page 687. Values specified directly in a graph template override style attributes.

When you are done with the NEWSTYLE style, you do not need to restore the STATISTICAL style template since you did not modify it. Rather, you inherited from the STATISTICAL style.

Changing the Default Style

The default style for each ODS destination is specified in the SAS Registry. For example, the default style for the HTML destination is DEFAULT and the default style for the RTF destination is RTF. You can specify a default style for all of your output in a particular ODS destination. This is useful if you want to use a different SAS style, if you have modified one of the styles supplied by SAS (see the section "Style Definitions and Colors" on page 667), or if you have defined your own style. For example, you can specify the JOURNAL style as the default style for RTF output.

The recommended approach for specifying a default style is as follows. Open the SAS Registry Editor by typing **regedit** in the command line. Expand the node **ODS ▶ DESTINATIONS** and select a destination (for example, select **RTF**). Double-click the **Selected Style** item, shown in Figure 21.31, and specify a style. This can be any style supplied by SAS or a user-defined style, as long as it can be found with the current template search path (for example, specify **Journal**). You can specify a default style for the other destinations in a similar way.

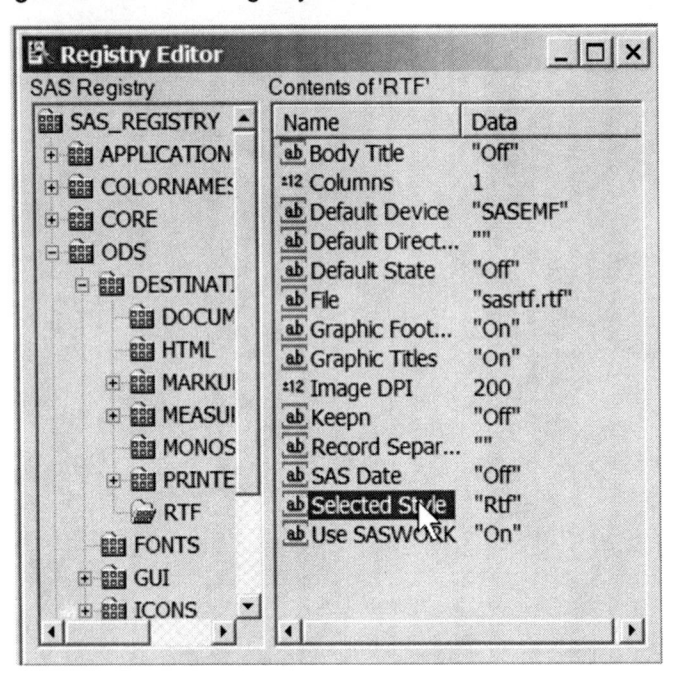

Figure 21.31 SAS Registry Editor

ODS searches sequentially through each element of the template search path for the first style definition that matches the name of the style specified in the SAS Registry. The first style definition found is used. (See the sections "Saving Customized Templates" on page 697, "Using Customized Templates" on page 697, and "Reverting to the Default Templates" on page 698 for more information about the template search path.) If you are specifying a customized style as your default style, the following are useful suggestions:

- If you save your style in SASUSER.Templat, verify that the name of your default style matches the name of the style specified in the SAS Registry. For example suppose the RTF style is specified for the RTF destination in the SAS Registry. You can name your style RTF and save it in SASUSER.Templat. This blocks the RTF style in SASHELP.Tmplmst (provided that you did not alter the default template search path).

- If you save your style in a user-defined template store, verify that this template store is the first in the current template search path. Include the ODS PATH statement in your SAS autoexec file so that it is executed at startup.

For the HTML destination, an alternative approach for specifying a default style is as follows. From the menu at the top of the main SAS window, select **Tools ▶ Options ▶ Preferences**. In the **Results** tab, check the **Create HTML** box and select a style from the pull-down menu.

Graph Templates

Graph templates control the layout and details of graphs produced with ODS Graphics. The SAS System provides a template for every graph produced by statistical procedures. Graph template definitions are written in the Graph Template Language (GTL). This powerful language includes statements for specifying plot layouts (such as lattices or overlays), plot types (such as scatter plots and histograms), and text elements (such as titles, footnotes, and insets). It also provides support for built-in computations (such as histogram binning) and the evaluation of expressions. Options are available for specifying colors, marker symbols, and other attributes of plot features.

Graphs, like all SAS output, are constructed from two underlying components, a data component (or data object) and a template. Procedures supply a table of data values and statistical results to plot. Together, the data object and the template form an output object that ODS displays in one or more output destinations. You can control this display in two ways. You can use the ODS Graphics Editor (discussed in the section "ODS Graphics Editor" on page 658) to modify the output object (but not the underlying data object or template), and you can use the GTL to modify the template. With just a little knowledge of the GTL, you can modify or edit templates, even when you do not understand most of the syntax used in the template definition. See examples starting with Example 21.5.

NOTE: You do not need to know anything about the GTL to create statistical graphics.

This section provides an overview of the Graph Template Language. It also describes how to locate, display, edit, and save templates. A *template definition* is a set of SAS statements that is used together with PROC TEMPLATE to create a compiled template. In addition to graph templates, two other common types of templates are table templates and style templates. A table template describes how to display the output for an output object that is rendered as a table. A style template provides formatting information for visual aspects of your SAS output, including both tables and graphs. In most applications, you do not have to modify the templates that are supplied by SAS. However, when customization is necessary, you can modify the default template with the template language and PROC TEMPLATE.

Compiled templates are stored in a template store, which is a type of item store. (An item store is a special type of SAS file.) The default templates supplied by SAS are stored in the SASHELP.Tmplmst template store. If you are using the SAS windowing environment, an easy way to display, edit, and save your templates is by using the Templates window. For detailed information about managing templates, see the *SAS Output Delivery System: User's Guide* and the *SAS/GRAPH: Graph Template Language User's Guide*. For details about the syntax of the graph template language, see the *SAS/GRAPH: Graph Template Language Reference*.

The Graph Template Language

Graph template definitions begin with a DEFINE STATGRAPH statement in PROC TEMPLATE, and they end with an END statement. Embedded in every graph template is a BEGINGRAPH/ENDGRAPH block, and embedded in that block are one or more LAYOUT blocks. You can specify the DYNAMIC statement to define dynamic variables (which the procedure uses to pass values to the template definition), the MVAR and NMVAR statements to define macro variables (which you can use to pass values to the template definition), and the NOTES statement to provide descriptive information about the graph. The default templates supplied by SAS for statistical procedures are often lengthy and complex, because they provide ODS Graphics with comprehensive and detailed information about graph construction. Here is one of the simpler graph templates for a statistical procedure:

```
define statgraph Stat.MDS.Graphics.Fit;
   notes "MDS Fit Plot";
   dynamic head;
   begingraph / designwidth=defaultdesignheight;
      entrytitle HEAD;
      layout overlayequated / equatetype=square;
         scatterplot y=FITDATA x=FITDIST / markerattrs=(size=5px);
         lineparm slope=1 x=0 y=0 / extend=true lineattrs=GRAPHREFERENCE;
      endlayout;
   endgraph;
end;
```

This template, supplied for the MDS procedure, creates a scatter plot of two variables, FitData and FitDist, along with a diagonal reference line that passes through the origin. The plot is square and the axes are equated so that a centimeter on one axis represents the same data range as a centimeter on the other axis. The plot title is provided by the evaluation of the dynamic variable Head, which is set by the procedure. It is not unusual for this plot to contain hundreds or even thousands of points, so a five-pixel marker is specified, which is smaller than the seven-pixel marker used by default in most styles.

The statements available in the graph template language can be classified as follows:

- Control statements specify the conditional or iterative flow of control. By default, flow of control is sequential. In other words, each statement is used in the order in which it appears.

- Layout statements specify the arrangement of the components of the graph. Layout statements are arranged in blocks that begin with a LAYOUT statement and end with an ENDLAYOUT statement. The blocks can be nested. Within a layout block, there can be plot, text, and other statements that define one or more graph components. Options provide control for attributes of layouts and components.

- Plot statements specify a number of commonly used displays, including scatter plots, histograms, contour plots, surface plots, and box plots. Plot statements are always provided within a layout block. The plot statements include options to specify the data columns from the source objects that are used in the graph. For example, in the SCATTERPLOT statement, there are mandatory X= and Y= arguments that specify which data columns are used for the X (horizontal) and Y (vertical) axes in the plot. (In the preceding example, FitData and FitDist are the names of columns int the data object that PROC MDS creates for this graph.) There is also a GROUP= option that specifies a data column as an optional classification variable.

- Text statements specify the descriptions that accompany graphs. An entry is any textual description, including titles, footnotes, and legends; it can include symbols to identify graph elements.

The following statements display another of the simpler template definitions—the definition of the scatter plot available in PROC KDE (see Figure 45.6.1 in Chapter 45, "The KDE Procedure"):

```
proc template;
   define statgraph Stat.KDE.Graphics.ScatterPlot;
      dynamic _TITLE _DEPLABEL _DEPLABEL2;
      BeginGraph;
         EntryTitle _TITLE;
         layout Overlay;
            scatterplot x=X y=Y / markerattrs=GRAPHDATADEFAULT;
         EndLayout;
      EndGraph;
   end;
run;
```

Here, the PROC TEMPLATE and RUN statements have been added to show how you would compile the template if you wanted to modify it. The DEFINE STATGRAPH statement in PROC TEMPLATE begins the graph template definition, and the END statement ends the definition. The DYNAMIC statement defines three dynamic variables that PROC KDE sets at run time. The variable _Title provides the title of the graph. The variables _DepLabel and _DepLabel2 contain the names of the X- and Y-variables, respectively. If you were to modify this template, you could use these dynamic text variables in any text element of the graph definition.

The overall display is specified with the LAYOUT OVERLAY statement inside the BEGIN-GRAPH/ENDGRAPH block. The title of the graph is specified with the ENTRYTITLE statement. The main plot is a scatter plot specified with the SCATTERPLOT statement. The options in the SCATTERPLOT statement are given after the slash and specify display options such as marker attributes (symbol, color, and size). These attributes can be specified directly, as in the PROC MDS template, or more typically by using indirect references to style attributes, as in the PROC KDE template. The values of these attributes are specified in the definition of the style you are using and are automatically set to different values if you specify a different style. For more information about style references, see the section "Styles" on page 664. The ENDLAYOUT statement ends the main layout block. For details about the syntax of the graph template language, see the *SAS/GRAPH: Graph Template Language Reference*.

You can write your own templates and use them to display raw data or output from procedures. For example, consider the iris data from Example 31.1 of Chapter 31, "The DISCRIM Procedure." The following statements create the SAS data set:

```
proc format;
   value specname
      1='Setosa   '
      2='Versicolor'
      3='Virginica ';
run;

data iris;
   input SepalLength SepalWidth PetalLength PetalWidth
         Species @@;
   format Species specname.;
   label SepalLength='Sepal Length in mm.'
         SepalWidth ='Sepal Width in mm.'
         PetalLength='Petal Length in mm.'
         PetalWidth ='Petal Width in mm.';
   datalines;
50 33 14 02 1 64 28 56 22 3 65 28 46 15 2 67 31 56 24 3

   ... more lines ...

;
```

The following statements create a template for a scatter plot of the variables PetalLength and PetalWidth with a legend:

```
proc template;
   define statgraph scatter;
      begingraph;
         entrytitle 'Fisher (1936) Iris Data';
         layout overlayequated / equatetype=fit;
            scatterplot x=petallength y=petalwidth /
                        group=species name='iris';
            layout gridded / autoalign=(topleft);
               discretelegend 'iris' / border=false opaque=false;
            endlayout;
         endlayout;
      endgraph;
   end;
run;
```

The layout is OVERLAYEQUATED, which equates the plot. However, unlike the PROC MDS template, which used EQUATETYPE=SQUARE to make a square plot, the EQUATETYPE=FIT option specifies that the lengths of the axes in this plot should fill the entire plotting area. A legend is placed internally in the top-left portion of the plot. There are three groups of observations, indicated by the three species, and each group is plotted with a separate color and symbol that depends on the ODS style. The legend identifies each group. The NAME= option provides the link between the SCATTERPLOT statement and the DISCRETELEGEND statement. An explicit link is needed since some graphical displays are based on multiple SCATTERPLOT statements or other plotting statements.

The following step creates the plot by using the SGRENDER procedure, the Iris data set, and the custom template `scatter`:

```
proc sgrender data=iris template=scatter;
run;
```

The syntax of PROC SGRENDER is very simple, because all of the graphical options appear in the template. The scatter plot in Figure 21.32 shows the results.

Figure 21.32 Petal Width and Petal Length in Three Iris Species

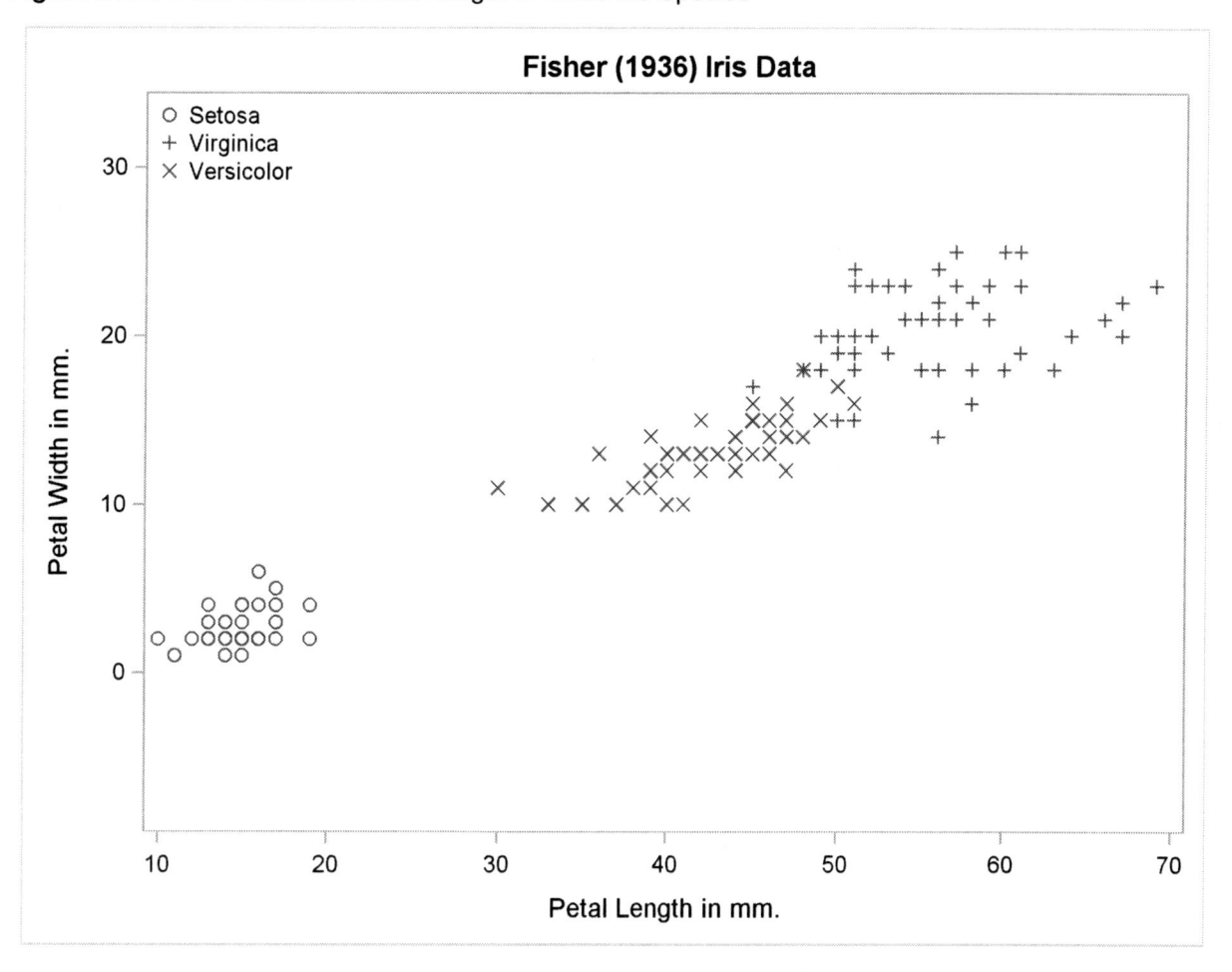

The intent of this example is to illustrate how you can write a template to create a scatterplot. PROC TEMPLATE and PROC SGRENDER provide you with the power to create highly customized displays. However, usually you can use the SGPLOT, SGSCATTER or SGPANEL procedures instead, which are much simpler to use. These procedures are discussed in section "Statistical Graphics Procedures" on page 708. See the section "Grouped Scatter Plot with PROC SGPLOT" on page 625 and Figure 21.12 for an example that plots these data with PROC SGPLOT.

Locating Templates

Before you can customize a graph, you must determine which template is used to create the original graph. You can do this by submitting the ODS TRACE ON statement before the procedure statements that create the graph. The fully qualified template name is displayed in the SAS log. Here is an example:

```
ods trace on;
ods graphics on;

proc reg data=sashelp.class;
   model Weight = Height;
run; quit;
```

The preceding statements create the following trace output, which provides information about both the graphs and tables produced by PROC REG:

```
Output Added:
-------------
Name:       NObs
Label:      Number of Observations
Template:   Stat.Reg.NObs
Path:       Reg.MODEL1.Fit.Weight.NObs
-------------

Output Added:
-------------
Name:       ANOVA
Label:      Analysis of Variance
Template:   Stat.REG.ANOVA
Path:       Reg.MODEL1.Fit.Weight.ANOVA
-------------

Output Added:
-------------
Name:       FitStatistics
Label:      Fit Statistics
Template:   Stat.REG.FitStatistics
Path:       Reg.MODEL1.Fit.Weight.FitStatistics
-------------

Output Added:
-------------
Name:       ParameterEstimates
Label:      Parameter Estimates
Template:   Stat.REG.ParameterEstimates
Path:       Reg.MODEL1.Fit.Weight.ParameterEstimates
-------------
```

```
Output Added:
-------------
Name:       DiagnosticsPanel
Label:      Fit Diagnostics
Template:   Stat.REG.Graphics.DiagnosticsPanel
Path:       Reg.MODEL1.ObswiseStats.Weight.DiagnosticPlots.DiagnosticsPanel
-------------

Output Added:
-------------
Name:       ResidualPlot
Label:      Height
Template:   Stat.REG.Graphics.ResidualPlot
Path:       Reg.MODEL1.ObswiseStats.Weight.ResidualPlots.ResidualPlot
-------------

Output Added:
-------------
Name:       FitPlot
Label:      Fit Plot
Template:   Stat.REG.Graphics.Fit
Path:       Reg.MODEL1.ObswiseStats.Weight.FitPlot
-------------
```

This is also illustrated in Example 21.5 and the section "The ODS Statement" on page 541 in Chapter 20, "Using the Output Delivery System."

Displaying Templates

Once you have found the fully qualified name of a template, you can display its definition (source program) by using one of these methods:

- Open the Templates window by issuing the command **odstemplates** (**odst** for short) in the command line of the SAS windowing environment. The template window is shown in Figure 21.33. If you expand the SASHELP.Tmplmst node, you can view all the available templates and double-click any template icon to display its definition. This is illustrated in Example 21.5.

- Use the SOURCE statement in PROC TEMPLATE to display a template definition in the SAS log or write the definition to a file.

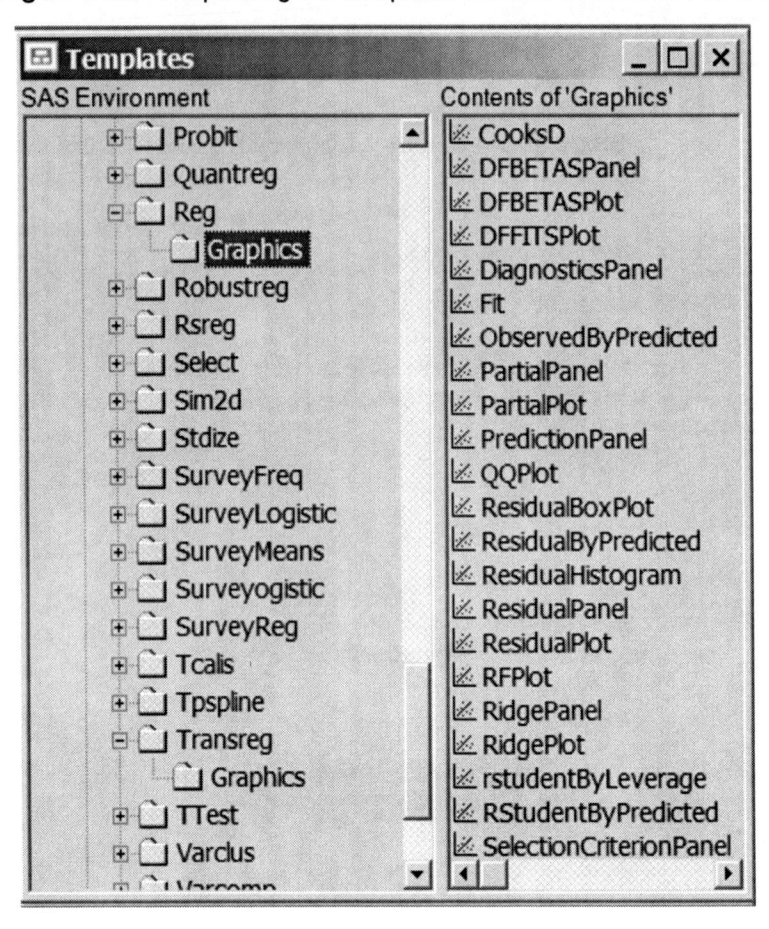

Figure 21.33 Requesting the Templates Window in the Command Line

For example, the following statements display the template for the PROC REG residual plot:

```
proc template;
   source Stat.REG.Graphics.ResidualPlot;
run;
```

The template is displayed as follows:

```
define statgraph Stat.Reg.Graphics.ResidualPlot;
   notes "Residual Plot";
   dynamic _XVAR _SHORTXLABEL _TITLE _LOESSLABEL _DEPNAME
         _MODELLABEL _SMOOTH;
BeginGraph;
   entrytitle halign=left textattrs=GRAPHVALUETEXT
      _MODELLABEL halign=center
      textattrs=GRAPHTITLETEXT _TITLE " for " _DEPNAME;
   entrytitle textattrs=GRAPHVALUETEXT _LOESSLABEL;
   layout overlay / xaxisopts=(shortlabel=_SHORTXLABEL);
      referenceline y=0;
      scatterplot y=RESIDUAL x=_XVAR / primary=true
         rolename=(_tip1=OBSERVATION _id1=ID1 _id2=ID2
         _id3=ID3 _id4=ID4 _id5=ID5) tip=(y x
         _tip1 _id1 _id2 _id3 _id4 _id5);
```

```
            if (EXISTS(_SMOOTH))
                loessplot y=_SMOOTH x=_XVAR /
                    tiplabel=(y="Smoothed Residual");
            endif;
        endlayout;
    EndGraph;
end;
```

PROC TEMPLATE also tells you where the template is located. In this case, it prints the following note:

```
NOTE: Path 'Stat.Reg.Graphics.ResidualPlot' is in: SASHELP.TMPLMST.
```

The word "Path" in ODS refers to any name or label hierarchy. In the note, the levels of the template name form a path. In the trace output, the levels of the plot name form a different path.

Editing Templates

You can modify the format and appearance of a particular graph by doing the following:

- Modify its template definition (source program).
- Submit the revised template to create a new compiled template.
- Ensure that the ODS search path finds and uses your new template.

Template stores are designated read-only (such as SASHELP.Tmplmst) or updatable (such as SASUSER.Templat).

If you view the templates in an updatable template store from the Templates window, you can select **Open** or **Edit** from the pop-up menu. Either the Template Browser or Template Editor window opens. In the Template Editor window, you can make changes and submit the code directly. For read-only templates or when you select **Open**, the Template Browser window opens and you must copy the definition to an editor window to make changes. Since templates supplied by SAS are in the read-only SASHELP library, an easy way to obtain an editable program file is to use the SOURCE statement with the FILE= option in PROC TEMPLATE to write the template definition to a file as follows:

```
proc template;
    source Stat.REG.Graphics.ResidualPlot / file="residtpl.sas";
run;
```

By default, the file is saved in the SAS current folder. Alternatively, you can omit the slash and the FILE= option and copy and paste the source from the SAS log into an editor. Either way, you must add a PROC TEMPLATE statement before the generated source statements and a RUN statement after the END statement before you submit your modified definition.

Graph definitions are self-contained and do not support inheritance or "parenting" as do table definitions. Consequently, the EDIT statement in PROC TEMPLATE is not supported for graph definitions.

Here are some important points about what you can and cannot change in a template supplied by SAS while preserving its overall functionality:

- Do not change the template name. A statistical procedure can access only a predefined list of templates. If you change the name, the procedure cannot find your template. You must keep the original name and make sure that it is in a template store that is read before SASHELP.Tmplmst. You control this with the ODS PATH statement (see the section "The Default Template Stores and the Template Search Path" on page 662, the section "Saving Customized Templates" on page 697, and subsequent sections for more information about the template search path and the ODS PATH statement).

- Do not change the names of columns. The underlying data object contains predefined column names that you must use. Be very careful if you change how a column is used in a template. Usually, columns are not interchangeable.

- Do not change the names of DYNAMIC variables. Procedures set values only for a predefined list of dynamic variables. Changing dynamic variable names can lead to runtime errors. Do not add dynamic variables, because the procedure cannot set their values. A few procedures document additional dynamic variables that can be defined in the template if you want to add more information to the output, such additional statistics in an inset table. See the section "Modifying the Layout and Adding a New Inset Table" on page 745 for an example.

- Do not change the names of statements (for example, from a SCATTERPLOT to a NEEDLE-PLOT or other type of plot).

You can change any of the following:

- You can add macro variables that behave like dynamic variables. They are resolved at the time that the statistical procedure is run, and not at the time that the template is compiled. They are defined with an MVAR or NMVAR statement at the beginning the template. You can set the value of each macro variable with a %LET statement before the statistical procedure is run. See Example 21.10. You can also move a variable from a DYNAMIC statement to an MVAR or NMVAR statement if you want to set it yourself rather than letting the procedure set it.

- You can change the graph size.

- You can change graph titles, footnotes, axis labels, and any other text that appears in the graph.

- You can change which plot features are displayed.

- You can change axis features, such as grid lines, offsets, view ports, tick value formatting, and so on.

- You can change the content and arrangement of insets (small tables of statistics embedded in some graphs).

- You can change the legend location, contents, border, background, title, and so on.

See the *SAS/GRAPH: Graph Template Language Reference* for information about the syntax of the statements in the Graph Template Language.

Saving Customized Templates

After you edit the template definition, you can submit your PROC TEMPLATE statements as you would any other SAS program. If you are using the Template Editor window, select **Submit** from the **Run** menu. See Example 21.5. Alternatively, submit your PROC TEMPLATE statements from the Program Editor. ODS automatically saves the compiled template in the first template store that it can update, according to the currently defined template search path. If you have not changed the template search path, then the modified template is saved in the SASUSER.Templat template store. You can display the current template search path with the following statement:

```
ods path show;
```

The log messages for the default template search path are as follows:

```
Current ODS PATH list is:

1. SASUSER.TEMPLAT(UPDATE)
2. SASHELP.TMPLMST(READ)
```

If you want to store modified templates in another template store, you can use the ODS PATH statement to add that template store to the front of the list. To use these templates, you must make sure the template search path is set correctly before you attempt to access them in the other SAS sessions. See the section "Using Customized Templates" on page 697.

Using Customized Templates

When you create ODS output (either graphs or tables), ODS searches sequentially through each template store in the template search path for a template that matches the one requested. If you have not changed the default template search path, then ODS searches the SASUSER.Templat store first, then SASHELP.Tmplmst. ODS uses the first template that it finds with the requested name. **NOTE:** Templates with the same name can exist in more than one template store.

The ODS PATH statement specifies the template stores to search, as well as the order in which to search them. You can change the default template search path by using the ODS PATH statement. For example, the following statement sets the template search path so that the template store WORK.Mystore is searched first, followed by SASHELP.Tmplmst:

```
ods path work.mystore(update) sashelp.tmplmst(read);
```

The UPDATE option provides update access as well as read access to WORK.Mystore. The READ option provides read-only access to SASHELP.Tmplmst. With this path, the template store SASUSER.Templat is no longer searched. You can verify this with the following statement:

```
ods path show;
```

The log messages generated by the preceding statement are as follows:

```
Current ODS PATH list is:

1. WORK.MYSTORE(UPDATE)
2. SASHELP.TMPLMST(READ)
```

For more information, see the *SAS Output Delivery System: User's Guide* and the *SAS/GRAPH: Graph Template Language User's Guide*. Example 21.5 illustrates all the steps of displaying, editing, saving, and using customized templates.

Reverting to the Default Templates

Customized templates are stored in SASUSER.Templat or in some other template store that you create. The templates supplied by SAS are in the read-only template store SASHELP.Tmplmst. If you have modified any of the supplied templates and you want to use the original default templates, you can change your template search path as follows:

```
ods path sashelp.tmplmst(read) sasuser.templat(update);
```

This way the default templates are found first. Alternatively, you can save all of your customized templates in a user-defined template store (for example WORK.Mystore). To access these templates, you submit the following statement before running your analysis:

```
ods path mylib.mystore(update) sashelp.tmplmst(read);
```

When you are done, you can reset the default template search path as follows:

```
ods path reset;
```

This restores the template search path to its original state (`sasuser.templat(update) sashelp.tmplmst(read)`). You can also save your customized template as part of your SAS program. You can delete it from the SASUSER.Templat template store when you are done, as in the following statements:

```
proc template;
   delete Stat.REG.Graphics.ResidualPlot;
run;
```

The following note is printed in the SAS log:

```
NOTE: 'Stat.REG.Graphics.ResidualPlot' has been deleted from: SASUSER.TEMPLAT
```

You can run the following step to delete the entire SASUSER.Templat store of customized templates:

```
ods path sashelp.tmplmst(read);
proc datasets library=sasuser nolist;
   delete templat(memtype=itemstor);
run;
ods path sasuser.templat(update) sashelp.tmplmst(read);
```

Template Modification Macros

The following two sections document two SAS autocall macros that you can use to modify templates. The %MODSTYLE macro in section "Style Template Modification Macro" on page 699 modifies style templates. The %MODTMPLT macro in section "Graph Template Modification Macro" on page 701 modifies graph templates.

You do not have to include autocall macros (for example, with a `%include` statement). You can call them directly once they are properly installed. If your site has installed the autocall libraries supplied by SAS and uses the standard configuration of SAS supplied software, you need to ensure that the SAS system option MAUTOSOURCE is in effect to begin using the autocall macros. For more information about autocall libraries, see the *SAS Macro Language: Reference*. For details about installing autocall macros, consult your host documentation.

Style Template Modification Macro

The %MODSTYLE macro provides easy ways to customize the style elements (`GraphData1`—`GraphDatan`) that control how groups of observations are distinguished. Examples of using the %MODSTYLE macro can be found in sections "Creating an All-Color Style by Using the ModStyle Macro" on page 675 and "Changing the Default Markers and Lines" on page 677. Also see Kuhfeld (2009) for more information about this macro.

The %MODSTYLE macro has the following options:

COLORS=_color-list_
> specifies a space-delimited list of colors for markers and lines. If you do not specify this option, then the colors from the parent style are used. You can specify the colors using any SAS color notation such as **cx**_rrggbb_.
>
> COLORS=GRAYS generates seven distinguishable grayscale colors from blackest to whitest. The colors should be mixed up to be more easily distinguished when you need fewer colors, but you can do that with your own COLORS= list. The HLS (hue/light/saturation) coding generates colors by setting hue and saturation to 0 and incrementing the lightness for each gray. You can also use the keywords BLUES, PURPLES, MAGENTAS, REDS, ORANGES, YELLOWS, GREENS, and CYANS to generate seven colors with a fixed hue and a saturation of AA (hex).
>
> COLORS=SHADES INT generates seven colors as described previously, except that you specify an integer $0 \leq INT < 360$. See *SAS/GRAPH: Reference*. The available hues include: GRAY, GREY, BLUE=0, PURPLE=30, MAGENTA=60, RED=120, ORANGE=150, YELLOW=180, GREEN=240, and CYAN=300.

DISPLAY=_n_
> specifies whether to display the generated template. By default, the template is not displayed. Specify DISPLAY=1 to display the generated template.

FILLCOLORS=*color-list*

specifies a space-delimited list of colors for bands and fills. If you do not specify this option, then the colors from the parent style are used.

Fill colors from the parent style are designed to work well with the colors from the parent style. If you specify a COLORS= list, then you might want to redefine the FILLCOLORS= list as well. You need to have at least as many fill colors as you have colors (any extra fill colors are ignored). Two shortcuts are available: FILLCOLORS=COLORS uses the COLORS= colors for the fills (your confidence bands should have transparency for this to be useful) and FILLCOLORS=LIGHTCOLORS modifies the lightness associated with each color generated by COLORS=SHADES (this is allowed only with COLORS=SHADES).

LINESTYLES=*line-style-list*

specifies a space-delimited list of line styles. The default is:

```
LineStyles=Solid MediumDash MediumDashShortDash LongDash
       DashDashDot LongDashShortDash DashDotDot Dash
       ShortDashDot MediumDashDotDot ShortDash
```

Line style numbers can range from 1 to 46. Some line styles have names associated with them. You can specify either the name or the number for the following number/name pairs: 1 Solid, 2 ShortDash, 4 MediumDash, 5 LongDash, 8 MediumDashShortDash, 14 DashDashDot, 15 DashDotDot, 20 Dash, 26 LongDashShortDash, 34 Dot, 35 ThinDot, 41 ShortDashDot, 42 MediumDashDotDot.

MARKERS=*marker-list*

specifies a space-delimited list of marker symbols. By default, `Markers=Circle Plus X Triangle Square Asterisk Diamond`. The available marker symbols are listed in *SAS/GRAPH: Graph Template Language Reference*. Two shortcuts are available: MARKERS=FILLED is an alias for the specification `Markers=CircleFilled TriangleFilled SquareFilled DiamondFilled StarFilled HomeDownFilled`, and MARKERS=EMPTY is an alias for the specification `Markers=Circle Triangle Square Diamond Star HomeDown`.

NAME=*style-name*

specifies the name of the new style that you are creating. This name is used when you specify the style in an ODS destination statement (for example, ODS HTML STYLE=*style-name*). The default is NAME=NEWSTYLE.

NUMBEROFGROUPS=*n*

specifies n, the number of `GraphData`n style elements to create. The `GraphData1`–`GraphData`n style elements contain n combinations of colors, markers, and line styles. By default, 32 combinations are created.

PARENT=*style-name*

specifies the parent style. The new style inherits most of its attributes from the parent style. The default is PARENT=DEFAULT (which is the default style for HTML).

TYPE=*type-specification*

specifies how your new style cycles through colors, markers, and line styles. The default is TYPE=LMbyC.

These first three methods work well with all plots, because cycling line styles and markers together ensures that both scatterplot markers and series plot lines are distinguishable:

CLM

cycles through colors, line styles, and markers simultaneously. The first group uses the first color, line style, and marker; the second group uses the second color, line style, and marker; and so on. This is the method used by the ODS Graphics styles.

LMbyC

fixes line style and marker, cycles through colors, and then moves to the next line style and marker. This is the default and creates a style where the first groups are distinguished entirely by color.

CbyLM

fixes color, cycles through line style and marker, and then moves to the next color. This option uses the smaller of the number of line styles or the number of markers when cycling within a color.

The following two methods might not work well with all plots:

CbyLbyM

fixes color and line style, then cycles through markers, increments line style, and then cycles through markers. After all line styles have been used, then this option moves to the next color and continues.

LbyMbyC

fixes line style and marker, then cycles through colors, increments marker, and then cycles through colors. After all markers have been used, then this option moves to the next line style and continues. This is closest to the legacy SAS/GRAPH method.

Graph Template Modification Macro

You can use the %MODTMPLT macro to insert BY line information, titles, and footnotes in ODS Graphics. You can also use it to remove titles and perform other template modifications. See Kuhfeld (2009) for more information about this macro.

The %MODTMPLT macro has the following options:

BY=*by-variable-list*

specifies the list of BY variables. Also see BYLIST=. When graphs are produced (by default or when the STEPS= value contains 'G'), you must specify the BY= option. Otherwise, when you are only modifying the template, you do not need to specify the BY= option.

BYLIST=_by-statement-list_

specifies the full syntax of the BY statement. You can specify a full BY statement syntax including the DESCENDING or NOTSORTED options. If only BY variables are needed, specify only BY=. If you also need options, then specify the BY variables in the BY= option and the full syntax in the BYLIST= option (for example, specify BY=A B and BYLIST=A DESCENDING B).

DATA=_SAS-data-set_

specifies the input SAS data set. If you do not specify the DATA= option, the macro uses the most recently created SAS data set.

FILE=_filename_

specifies the file in which to store the original templates. This is a temporary file. You can specify either a quoted file name or the name from a FILENAME statement that you provide before you call the macro. The default is _"template.txt"_.

OPTIONS=_options_

specifies one or more of the following options (case is ignored):

LOG

displays a note in the SAS log when each BY group has finished.

FIRST

adds the ENTRYTITLE or ENTRYFOOTNOTE statements as the first titles or footnotes. By default, the statements are added after the last titles or footnotes. Most graph templates provided by SAS do not use footnotes; so this option usually affects only entry titles.

NOQUOTES

specifies that the values of the system titles and footnotes are to be moved to the ENTRYTITLE or ENTRYFOOTNOTE statements without the outer quotation marks. With OPTIONS=NOQUOTES, you can specify options in the titles or footnotes in addition to the text. However, you must ensure that you quote the text that provides the actual title or footnote.

The following is an example of an ordinary footnote:

```
footnote "My Footer";
```

With this FOOTNOTE statement and without OPTIONS=NOQUOTES, the macro creates the following ENTRYFOOTNOTE statement:

```
entryfootnote "My Footer";
```

The following footnotes are used with OPTIONS=NOQUOTES:

```
footnote 'halign=left "My Footer"';
footnote2 '"My Second Footer"';
```

With these FOOTNOTE statements and OPTIONS=NOQUOTES, the macro creates the following ENTRYFOOTNOTE statements:

```
entryfootnote halign=left "My Footer";
entryfootnote "My Second Footer";
```

REPLACE

replaces the unconditionally added entry titles and entry footnotes in the templates (those that are not part of IF or ELSE statements) with the system titles and footnotes. The system titles and footnotes are those that are specified in the TITLE or FOOTNOTE statements. You can instead use the TITLES=*SAS-data-set* option to specify titles and footnotes with a data set. If OPTIONS=REPLACE is specified, then OPTIONS=TITLES is ignored.

SOURCE

displays the generated source code. By default, the template source code is not displayed.

TITLES

displays the system titles and footnotes with the graphs. The system titles and footnotes are those that are specified in the TITLE or FOOTNOTE statements. You can instead use the TITLES=*SAS-data-set* option to specify titles and footnotes with a data set. If you also specify OPTIONS=FIRST, the system titles and footnotes are inserted before the previously existing entry titles and entry footnotes in the templates. Otherwise, they are inserted at the end.

You can specify OPTIONS=TITLES or OPTIONS=REPLACE, or insert BY lines, or do both. If you do both, and you do not like where the BY line is inserted relative to your titles and footnotes, just specify OPTIONS=NOQUOTES and _ByLine0 to place the BY line wherever you choose. The following TITLE statements illustrate:

```
title1 '"My First Title"';
title2 '_byline0';
title3 '"My Last Title"';
```

Also, you can embed BY information in a title or a footnote, again with OPTIONS=NOQUOTES. For example:

```
title '"Spline Fit By Sex, " _byline0';
```

When _ByLine0 is specified in any of the titles or footnotes, then the usual BY line is not added.

The following example removes all titles and footnotes:

```
footnote;
title;
%modtmplt(options=replace, template=Stat.Transreg.Graphics, steps=t)
```

STATEMENT=*entry-statement-fragment*

specifies the statement that contains the BY line that gets added to the template along with any statement options. The default is `Statement=EntryFootNote halign=left TextAttrs=GraphValueText`. Other examples include:

```
Statement=EntryTitle
Statement=EntryFootNote halign=left TextAttrs=GraphLabelText
```

STEPS=*steps*

specifies the macro steps to run. Case and white space are ignored. the macro modifies the templates (when 'T' is specified), produces the graphs for each BY group (when 'G' is specified), and deletes the modified templates (when 'D' is specified). The default is STEPS=TGD. You can instead have it perform a subset of these three tasks by specifying a subset of terms in the STEPS= option.

When you use the %MODTMPLT macro to add BY lines, you usually do not need to delete the templates before you run your procedure again in the normal way. The template modification inserts the BY line through a macro variable and an MVAR statement. When the macro variable _ByLine0 is undefined, the ENTRYTITLE or ENTRYFOOTNOTE statement drops out as if it were not there at all.

STMTOPTS1= *n* ADD | REPLACE | DELETE | BEFORE | AFTER *statement-name* <*options*>
STMTOPTS2= *n* ADD | REPLACE | DELETE | BEFORE | AFTER *statement-name* <*options*>
STMTOPTS3= *n* ADD | REPLACE | DELETE | BEFORE | AFTER *statement-name* <*options*>
STMTOPTS4= *n* ADD | REPLACE | DELETE | BEFORE | AFTER *statement-name* <*options*>
STMTOPTS5= *n* ADD | REPLACE | DELETE | BEFORE | AFTER *statement-name* <*options*>
STMTOPTS6= *n* ADD | REPLACE | DELETE | BEFORE | AFTER *statement-name* <*options*>
STMTOPTS7= *n* ADD | REPLACE | DELETE | BEFORE | AFTER *statement-name* <*options*>
STMTOPTS8= *n* ADD | REPLACE | DELETE | BEFORE | AFTER *statement-name* <*options*>
STMTOPTS9= *n* ADD | REPLACE | DELETE | BEFORE | AFTER *statement-name* <*options*>
STMTOPTS10=*n* ADD | REPLACE | DELETE | BEFORE | AFTER *statement-name* <*options*>

These ten options add or replace options in up to 10 selected statements. The following example illustrates:

```
%modtmplt(template=Stat.glm.graphics.residualhistogram, steps=t,
         stmtopts1=.  add  discretelegend  autoalign=(topleft),
         stmtopts2=1  add  densityplot     legendlabel='Normal Density',
         stmtopts3=2  add  densityplot     legendlabel='Kernel Density',
         stmtopts4=1  add  overlay         yaxisopts=(griddisplay=on)
                                           yaxisopts=(label='Normal and Kernel Density'))

proc glm plots=diagnostics(unpack) data=sashelp.class;
   model weight = height;
run;

%modtmplt(template=Stat.glm.graphics.residualhistogram, steps=d)
```

These options require you to specify a series of values. The first value is the statement number (or missing to modify options on all statements that match the statement name). The second value is: ADD, REPLACE, DELETE, BEFORE, or AFTER. When the second value is ADD or REPLACE, it controls whether you add new options or replace existing options. Alternatively, the second value can be BEFORE or AFTER to add a new statement before or after the named statement. When the value is DELETE, the corresponding statement is deleted. The third value is a statement name. All remaining options are options for the statement named by the third value (with ADD and REPLACE) or for a new statement (with BEFORE and AFTER). In the STMTOPTS1= example, an option is added to all DiscreteLegend statements. In the STMTOPTS2= example, an option is added to the first DensityPlot statement. In the STMTOPTS4= example, an option is added to the LAYOUT OVERLAY statement. In most cases, the statement name is the first name that begins the statement. The LAYOUT statement is an exception. In the case of layouts, specify the second name (OVERLAY, GRIDDED, LATTICE, and so on) for the third value. Note that a statement such as `if (expression) EntryTitle...;` is an IF statement not an ENTRYTITLE statement.

If an option is specified multiple times on a GTL statement, the last specification overrides previous specifications. Hence, you do not need to know and respecify all of the options. You can just add an option to the end, and it overrides the previous value. You can use these options only to modify statements that contain a slash, and only to modify the options that come after the slash. Note that in STMTOPTS4=, the YAXISOPTS= option is specified twice. It could have been equivalently specified once as follows:

```
yaxisopts=(griddisplay=on label='Normal and Kernel Density'))
```

The actual specification adds the GRIDDISPLAY=ON to the Y axis options (which by default has only a label specification). The old label is unchanged until the LABEL= option in the second YAXISOPTS= specification overrides it. In other words, YAXISOPTS=(GRIDDISPLAY=ON) augments the old YAXISOPTS= option; it does not replace it.

The following steps delete the legend and instead provide a footnote:

```
%modtmplt (template=Stat.glm.graphics.residualhistogram, steps=t,
          stmtopts1=. delete discretelegend,
          stmtopts2=1 after  begingraph entryfootnote
                      textattrs=GraphLabelText(color=cx445694) 'Normal '
                      textattrs=GraphLabelText(color=cxA23A2E) 'Kernel')

proc glm plots=diagnostics(unpack) data=sashelp.class;
   model weight = height;
run;

%modtmplt (template=Stat.glm.graphics.residualhistogram, steps=d)
```

TEMPLATE=SAS-template

specifies the name of the template to modify. You can specify just the first few levels to modify a series of templates. For example, to modify all of PROC REG's graph templates, specify TEMPLATE=`Stat.Reg.Graphics`. This option is required.

TITLES=SAS-data-set

> specifies a data set that contains titles or footnotes or both. By default, when the system titles or footnotes are used (when OPTIONS=TITLES or OPTIONS=REPLACE is specified), PROC SQL is used to determine the titles and footnotes. You can instead create this data set yourself so that you can set the graph titles independently from the system titles and footnotes. The data set must contain two variables: Type (Type='T' for titles and Type='F' for footnotes), and Text, which contains the titles and footnotes. Other variables are ignored. Specify the titles and footnotes in the order in which you want them to appear.

TITLEOPTS=entry-statement-options

> specifies the options for system titles and footnotes. For example, you can specify the HALIGN= and TEXTATTRS= options as in the STATEMENT= option. By default, no title options are used. With OPTIONS=NOQUOTES, you can specify options individually.

Adding a BY Line to Graphs

You can use the %MODTMPLT macro to display in your graphs BY line information (such as `Sex = 'F'` and `Sex = 'M'` when the statement BY SEX is specified). The %MODTMPLT macro requires you to construct a SAS macro called %MYGRAPH, which contains the SAS procedure that needs to be run, so that the %MODTMPLT macro can call it for each BY group. The following example illustrates this usage of the macro:

```
proc sort data=sashelp.class out=class;
   by sex;
run;

%macro mygraph;
proc transreg data=__bydata;
   model identity(weight) = spline(height);
%mend;

%modtmplt(by=sex, data=class, template=Stat.Transreg.Graphics)
```

Notice that the BY and RUN statements are *not* specified in the %MYGRAPH macro. Also notice that you must specify DATA=__BYDATA with the procedure call in the %MYGRAPH macro and specify the real input data set in the DATA= option of the %MODTMPLT macro.

The %MODTMPLT macro outputs the specified template or templates to a file, adds an ENTRYFOOTNOTE statement with the BY line information, and then runs the %MYGRAPH macro once for each BY group. In the end, the %MODTMPLT macro deletes the modified template.

The results of the preceding statements are displayed in Figure 21.34 and Figure 21.35. The BY line is displayed as a left-justified footnote by default. You can change this with the STATEMENT= option (default: `Statement=EntryFootNote Halign=Left TextAttrs=GraphValueText`). For example, you can display the BY line as a centered title by specifying STATEMENT=ENTRYTITLE.

Figure 21.34 The First BY Group

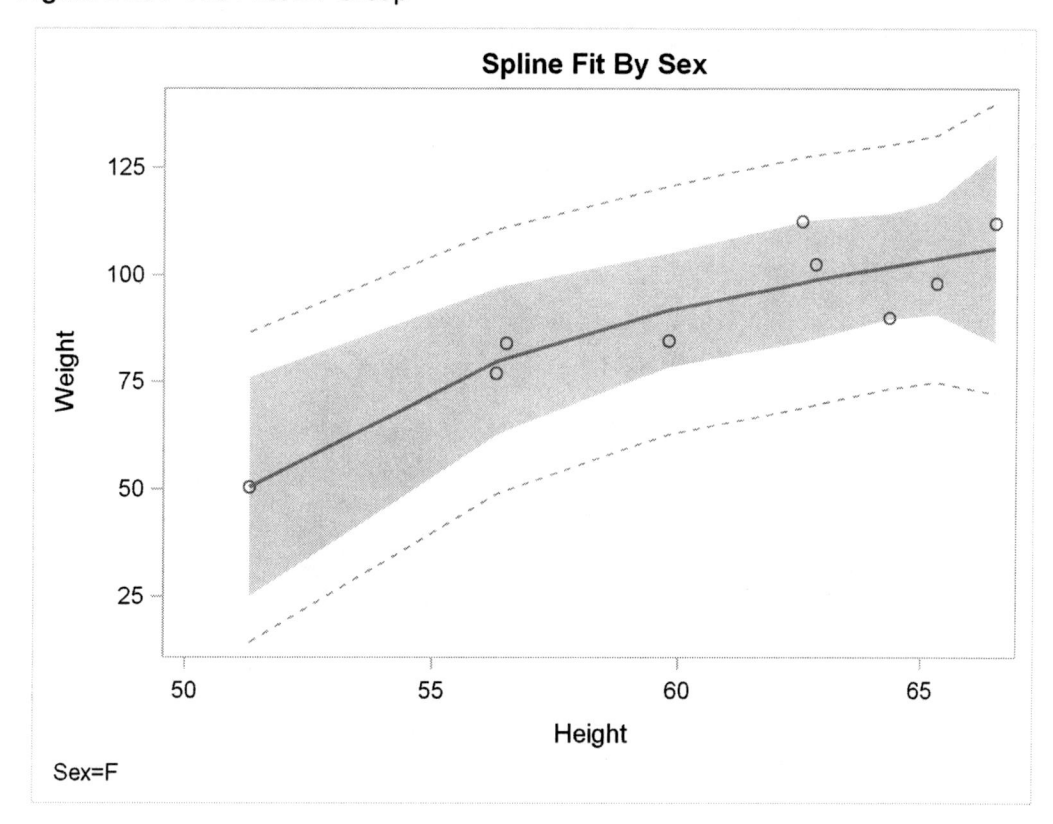

Figure 21.35 The Second BY Group

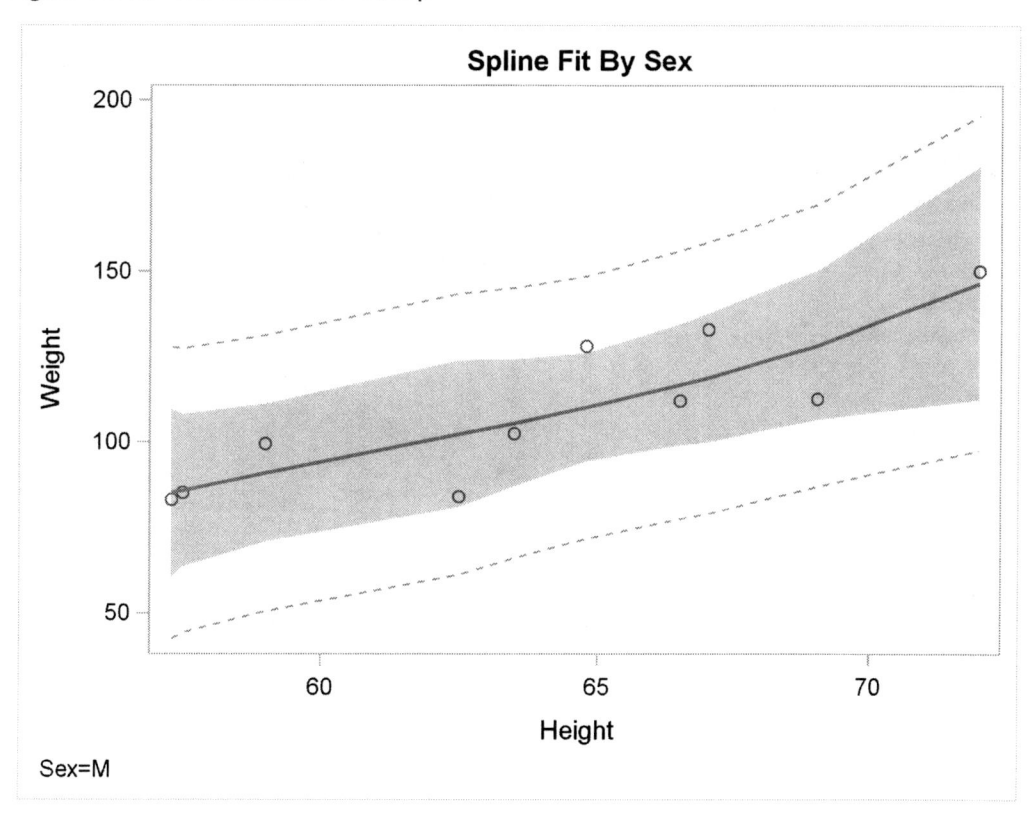

Statistical Graphics Procedures

There are three statistical graphics procedures in SAS/GRAPH that use ODS Graphics and provide convenient syntax for creating a variety of plots from raw data or from procedure output.

SGSCATTER   creates single-cell and multi-cell scatter plots and scatter plot matrices with optional fits and ellipses.

SGPLOT   creates single-cell plots with a variety of plot and chart types.

SGPANEL   creates single-page or multi-page panels of plots and charts conditional on classification variables.

In addition, the SGRENDER procedure in SAS/GRAPH provides a way to create plots from graph templates that you have modified or written yourself.

See the *SAS/GRAPH: Statistical Graphics Procedures Guide* for more information. You do not need to enable ODS Graphics in order to use the SG procedures.

These procedures do much more than make scatter plots. They can produce density plots, dot plots, needle plots, series plots, horizontal and vertical bar charts, histograms, and box plots. They can also compute and display loess fits, polynomial fits, penalized B-spline fits, reference lines, bands, and ellipses. PROC SGRENDER is the most flexible because it uses the Graph Template Language. The syntax for the other SG procedures is much simpler than that of the GTL, and so these procedures are recommended for creating most plots commonly required in statistical work.

The SGPLOT Procedure

PROC SGPLOT provides a simple way to make a variety of scatter plots. This example is taken from Example 50.4 of Chapter 50, "The LOESS Procedure." The following statements create a SAS data set that contains information about differences in ocean pressure over time:

```
data ENSO;
   input Pressure @@;
   Year = _n_ / 12;
   datalines;
12.9  11.3  10.6  11.2  10.9   7.5   7.7  11.7

   ... more lines ...

;
```

The following statements create a scatter plot of points along with a penalized B-spline fit to the data and produce Figure 21.36:

```
proc sgplot data=enso noautolegend;
   title 'Atmospheric Pressure Differences Between '
         'Easter Island and Darwin, Australia';
   pbspline y=pressure x=year;
run;
```

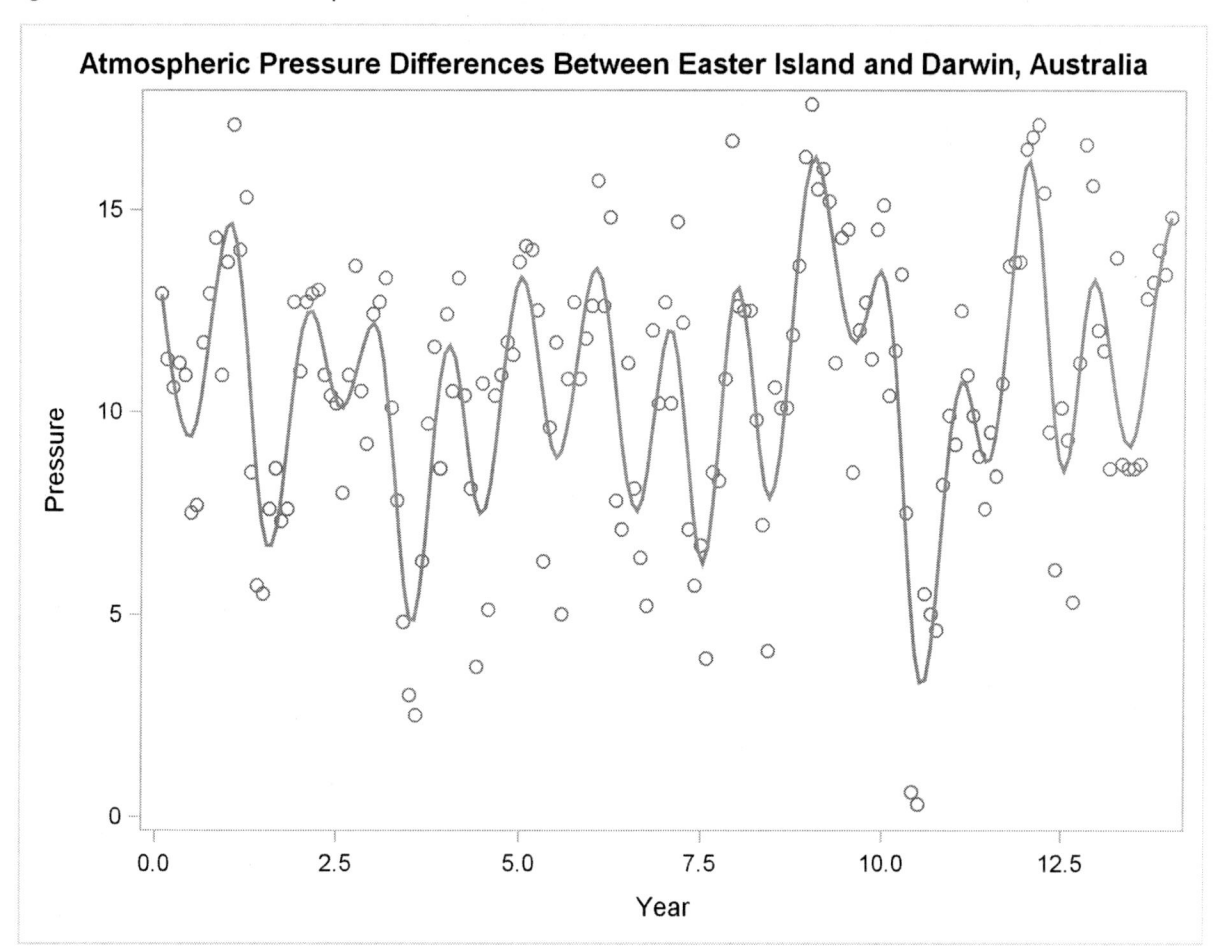

Figure 21.36 Penalized B-Spline Fit with PROC SGPLOT

[Figure: Scatter plot titled "Atmospheric Pressure Differences Between Easter Island and Darwin, Australia" with Pressure on y-axis (0 to 15+) and Year on x-axis (0.0 to 12.5), showing data points with a fitted spline curve.]

See Chapter 91, "The TRANSREG Procedure," for more information about penalized B-splines. Also see the section "Grouped Scatter Plot with PROC SGPLOT" on page 625 and Figure 21.12 for an example of a scatter plot with groups of observations.

The SGSCATTER Procedure

You can use the SGSCATTER procedure to produce scatter plot matrices. The following step creates a scatter plot matrix from all of the numeric variables in the Class data set available in the SASHELP library and produces Figure 21.37:

```
proc sgscatter data=sashelp.class;
   matrix _numeric_ / diagonal=(kernel histogram);
run;
```

The diagonal cells of Figure 21.37 contain a histogram and a kernel density fit. The off-diagonal cells contain all pairs of scatter plots.

Figure 21.37 Scatter Plot Matrix with PROC SGSCATTER

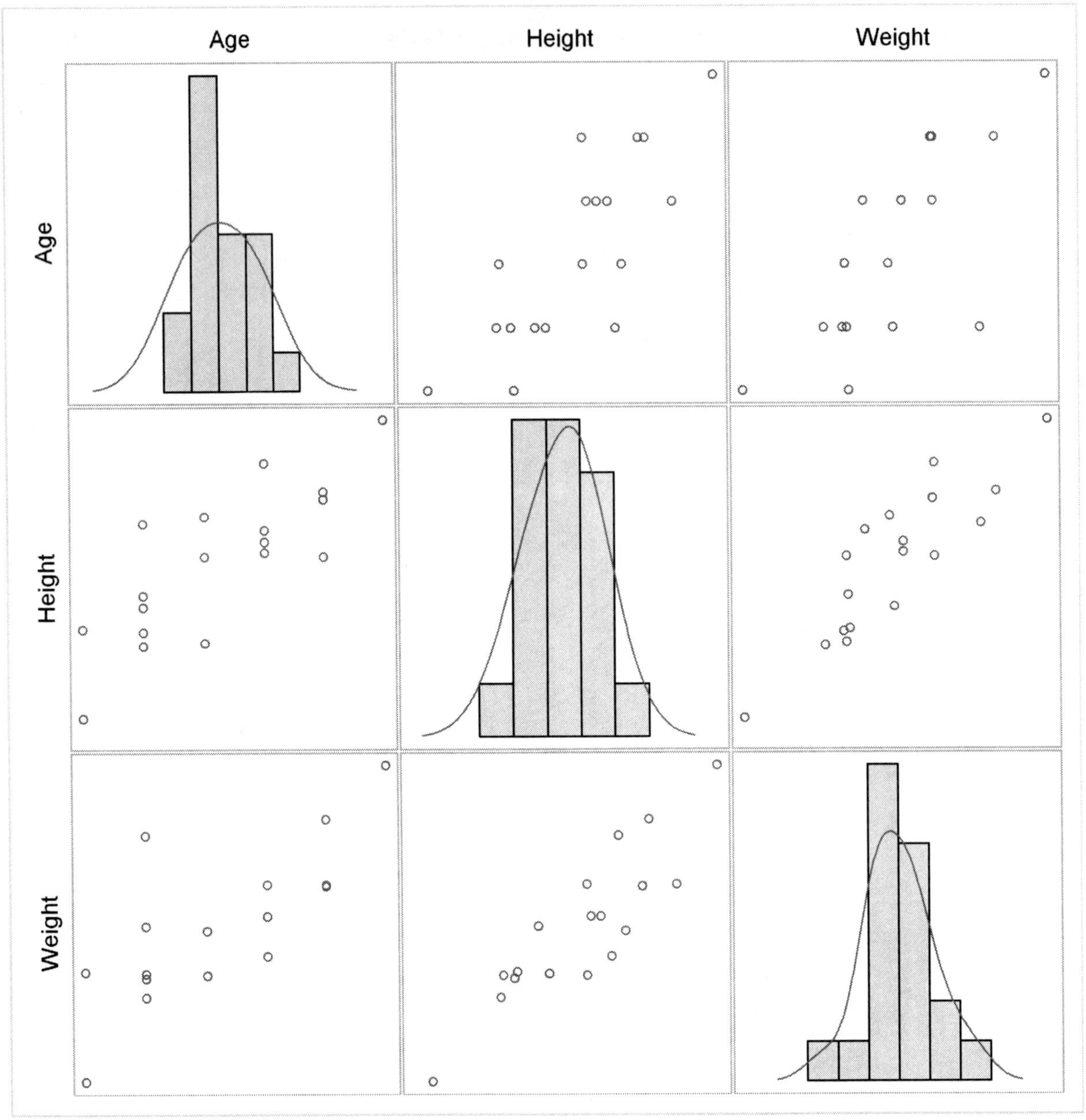

The MATRIX statement creates a symmetric $n \times n$ scatter plot matrix. Other statements are available as well. The PLOT statement creates a panel containing one or more individual scatter plots. The COMPARE statement creates a rectangular $m \times n$ scatter plot matrix. Linear and nonlinear fits can be added, and many graphical features can be requested with options.

The SGPANEL Procedure

The SGPANEL procedure creates paneled plots and charts with one or more classification variables. Classification variables can be designated as row or column variables, or there can be multiple classifications. Graphs are drawn for each combination of the levels of classification variables, showing a subset of the data in each cell.

This example is taken from Example 38.6 of Chapter 38, "The GLIMMIX Procedure." The following statements create the input SAS data sets:

```
data times;
   input time1-time23;
   datalines;
122  150  166  179  219  247  276  296  324  354  380  445
478  508  536  569  599  627  655  668  723  751  781
;

data cows;
   if _n_ = 1 then merge times;
   array t{23} time1 - time23;
   array w{23} weight1 - weight23;
   input cow iron infection weight1-weight23 @@;
   do i=1 to 23;
      weight = w{i};
      tpoint = (t{i}-t{1})/10;
      output;
   end;
   keep cow iron infection tpoint weight;
   datalines;
1 0 0   4.7    4.905  5.011  5.075  5.136  5.165  5.298  5.323
        5.416  5.438  5.541  5.652  5.687  5.737  5.814  5.799
        5.784  5.844  5.886  5.914  5.979  5.927  5.94

... more lines ...

;
```

First, PROC GLIMMIX is run to fit the model, and then the results are prepared for plotting:

```
proc glimmix data=cows;
  t2 = tpoint / 100;
  class cow iron infection;
  model weight = iron infection iron*infection tpoint;
  random t2 / type=rsmooth subject=cow
                  knotmethod=kdtree(bucket=100 knotinfo);
  output out=gmxout pred(blup)=pred;
  nloptions tech=newrap;
run;

data plot;
  set gmxout;
  length Group $ 26;
       if (iron=0) and (infection=0) then group='Control Group (n=4)';
  else if (iron=1) and (infection=0) then group='Iron - No Infection (n=3)';
  else if (iron=0) and (infection=1) then group='No Iron - Infection (n=9)';
  else group = 'Iron - Infection (n=10)';
run;

proc sort data=plot; by group cow;
run;
```

The following statements produce graphs of the observed data and fitted profiles in the four groups:

```
proc sgpanel data=plot noautolegend;
  title 'Radial Smoothing with Cow-Specific Trends';
  label tpoint='Time' weight='log(Weight)';
  panelby group / columns=2 rows=2;
  scatter x=tpoint y=weight;
  series  x=tpoint y=pred / group=cow lineattrs=GraphFit;
run;
```

The results are shown in Figure 21.38.

Figure 21.38 Fit Using PROC SGPANEL

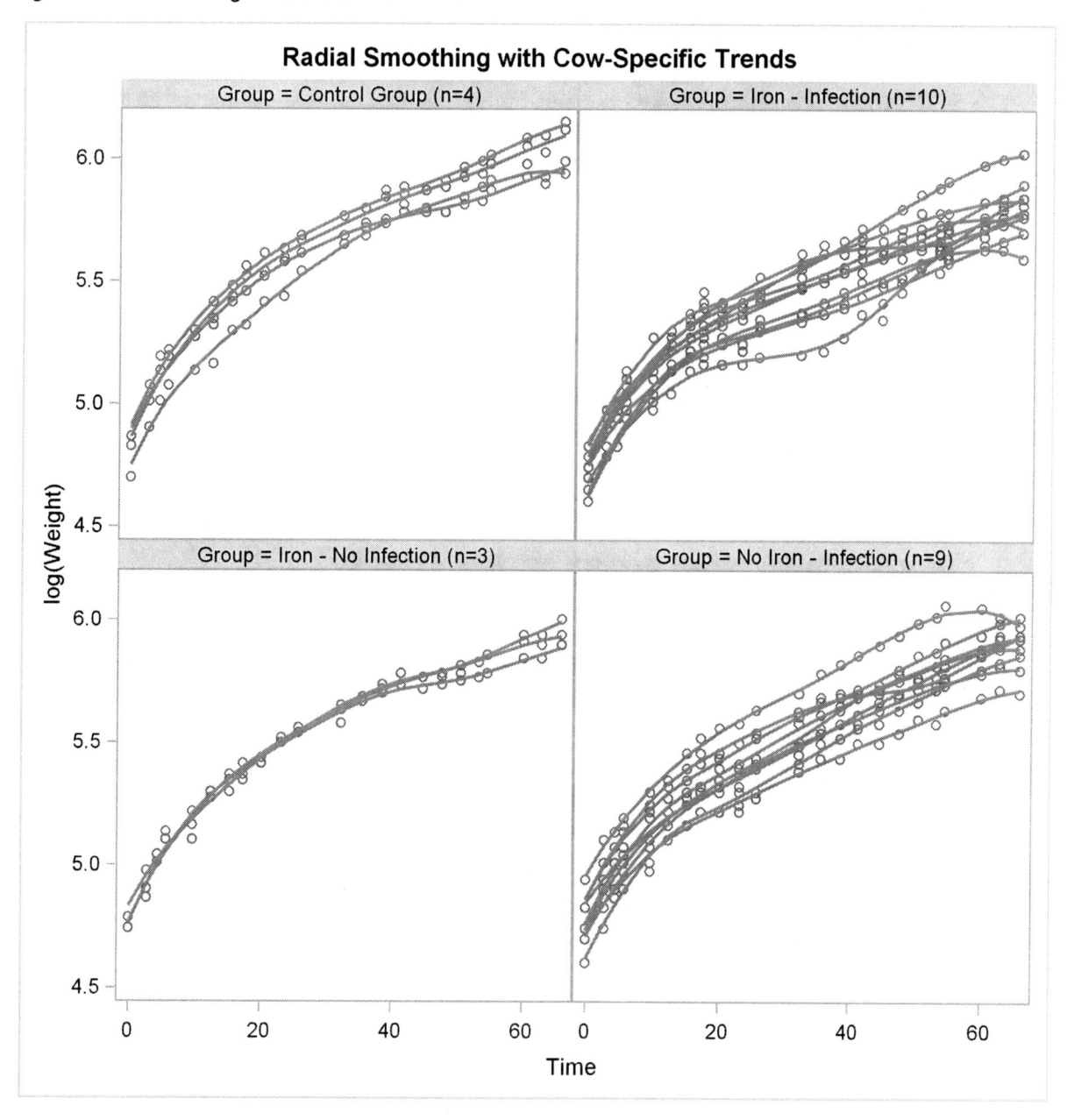

The SGRENDER Procedure

The SGRENDER procedure produces a graph from an input SAS data set and an ODS graph template. With PROC SGRENDER and the Graph Template Language (GTL), you can create highly customized graphs. The following steps create a simple scatter plot of the Class data set available in the SASHELP library and produce Figure 21.39:

```
proc template;
   define statgraph Scatter;
      begingraph;
         entrytitle "Simple Scatter Plot of the Class Data Set";
         layout overlay;
            scatterplot y=weight x=height / datalabel=name;
         endlayout;
      endgraph;
   end;
run;

proc sgrender data=sashelp.class template=scatter;
run;
```

The template definition consists of an outer block that begins with a DEFINE statement and ends with an END statement. Inside of that is a BEGINGRAPH/ENDGRAPH block. Inside that block, the ENTRYTITLE statement provides the plot title, and the LAYOUT OVERLAY block contains the statement or statements that define the graph. In this case, there is just a single SCATTERPLOT statement that names the Y-axis (vertical) variable, the X-axis (horizontal) variable, and an optional variable that contains labels for the points. The PROC SGRENDER statement simply specifies the input data set and the template. The real work in using PROC SGRENDER is writing the template.

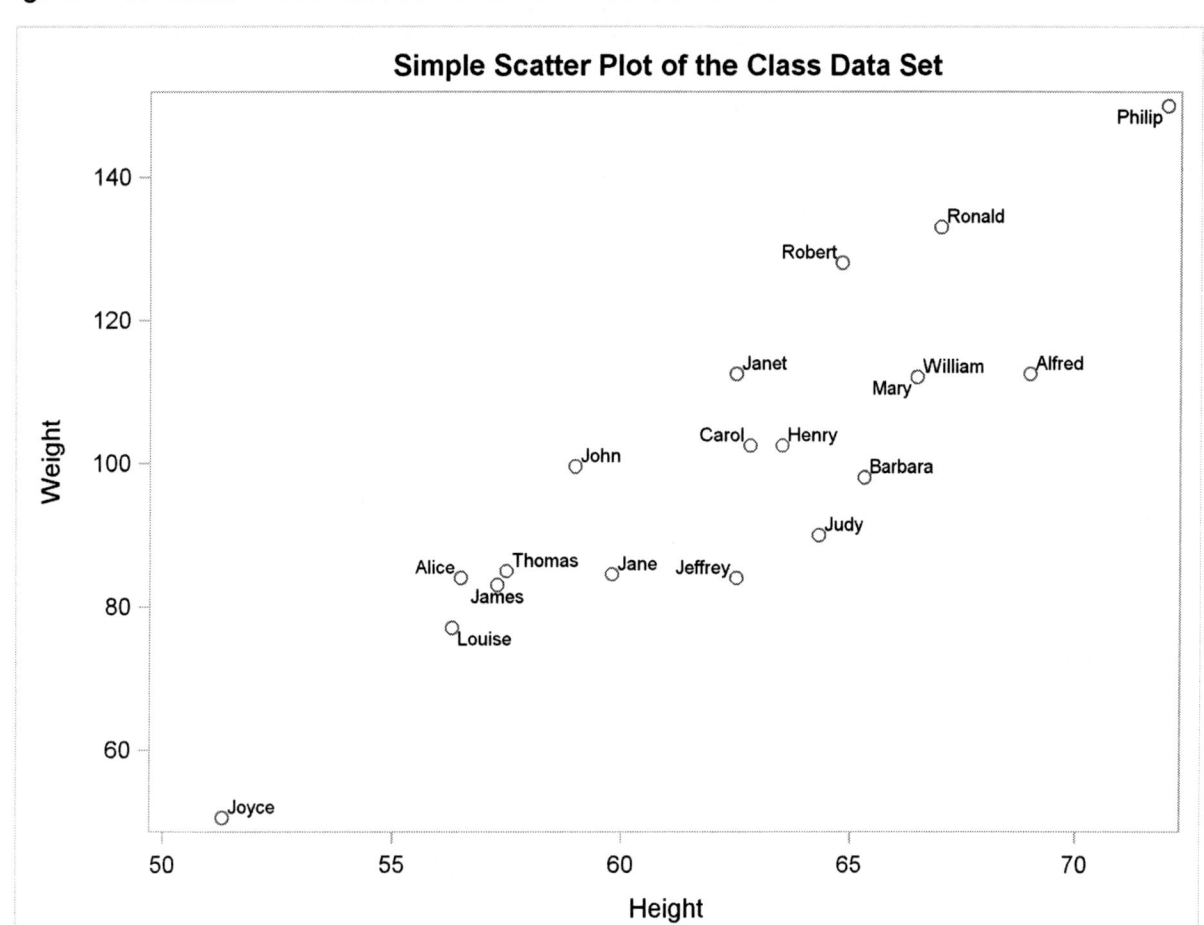

Figure 21.39 Scatter Plot of Labeled Points with PROC SGRENDER

The following steps add a series of fit functions to the scatter plot and create a legend by adding statements to the `Scatter` template:

```
proc template;
   define statgraph Scatter;
      begingraph;
         entrytitle "Scatter Plot of the Class Data Set with Fit Functions";
         layout overlay;
            scatterplot y=weight x=height / datalabel=name;
            pbsplineplot y=weight x=height / name='pbs'
                         legendlabel='Penalized B-Spline'
                         lineattrs=GraphData1;
            regressionplot y=weight x=height / degree=1 name='line'
                         legendlabel='Linear Fit'
                         lineattrs=GraphData2;
            regressionplot y=weight x=height / degree=3 name='cubic'
                         legendlabel='Cubic Fit'
                         lineattrs=GraphData3;
            loessplot y=weight x=height / name='loess'
                         legendlabel='Loess Fit'
                         lineattrs=GraphData4;
```

```
                    discretelegend 'pbs' 'line' 'cubic' 'loess';
            endlayout;
        endgraph;
    end;
run;

proc sgrender data=sashelp.class template=scatter;
run;
```

The line attributes for each function are specified with different style elements, **GraphData1** through **GraphData4**, so that the functions are adequately identified in the legend. The preceding statements create Figure 21.40.

Figure 21.40 Scatter Plot and Fit Functions with PROC SGRENDER

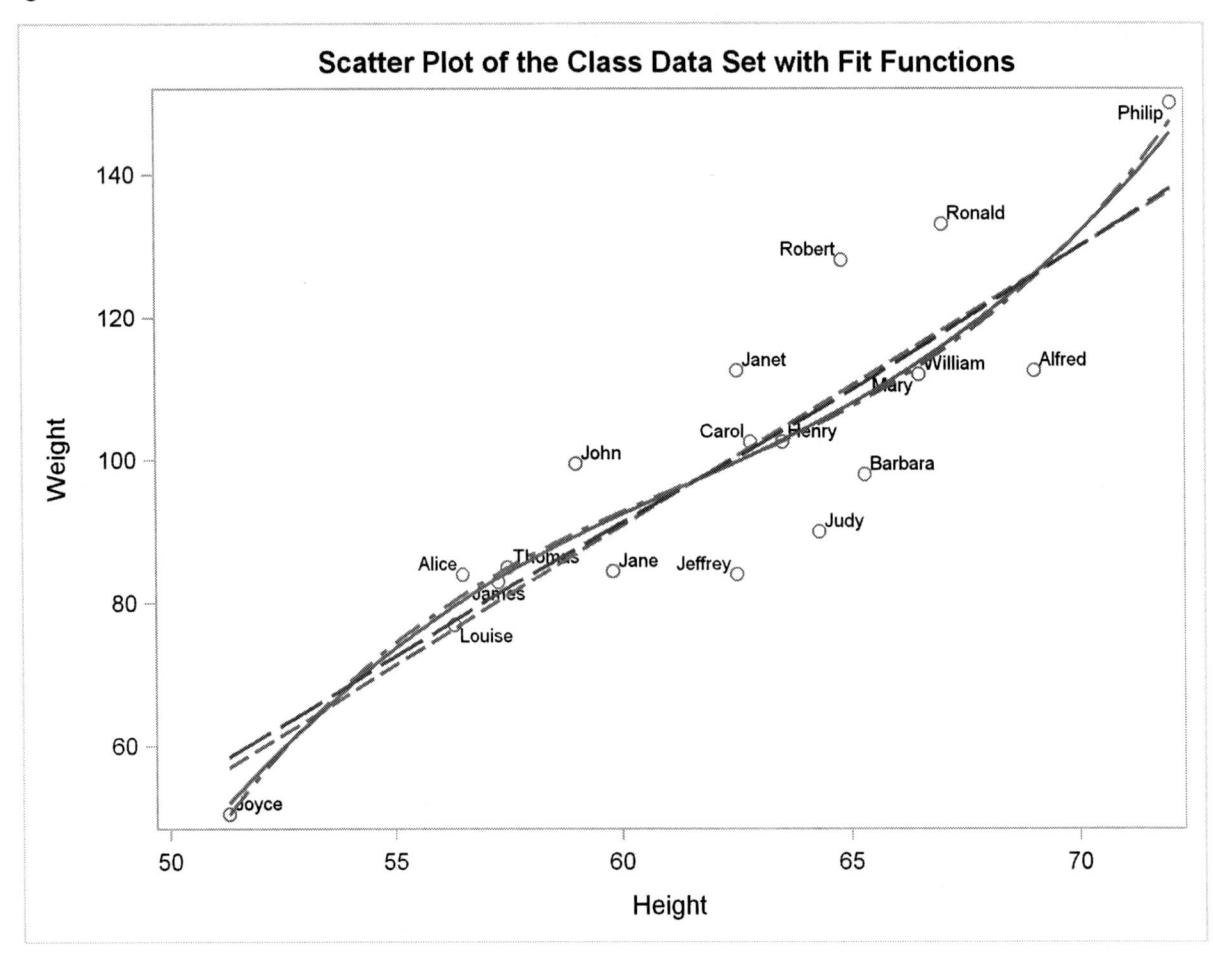

The following statements create a four-panel display of the Class data set and produce Figure 21.41:

```
proc template;
   define statgraph Panel;
      begingraph;
         entrytitle "Paneled Display of the Class Data Set";

         layout lattice / rows=2 columns=2 rowgutter=10 columngutter=10;

            layout overlay;
               scatterplot y=weight x=height;
               pbsplineplot y=weight x=height;
            endlayout;

            layout overlay / xaxisopts=(label='Weight');
               histogram weight;
            endlayout;

            layout overlay / yaxisopts=(label='Height');
               boxplot y=height;
            endlayout;

            layout overlay / xaxisopts=(offsetmin=0.1 offsetmax=0.1)
                             yaxisopts=(offsetmin=0.1 offsetmax=0.1);
               scatterplot  y=weight x=height / markercharacter=sex
                  name='color' markercolorgradient=age;
               continuouslegend 'color'/ title='Age';
            endlayout;

         endlayout;
      endgraph;
   end;
run;

proc sgrender data=sashelp.class template=panel;
run;
```

In this template, the outermost layout is a LAYOUT LATTICE. It creates a 2 × 2 panel of plots with a ten-pixel separation (or gutter) between each plot. Inside the lattice are four LAYOUT OVERLAY blocks—each defining one of the graphs. The first is a simple scatter plot with a nonlinear penalized B-spline fit. The second is a histogram of the dependent variable Weight. The third is a box plot of the independent variable Height. The fourth simultaneously shows height, weight, age, and sex for the students in the class. Each axis has an offset added at both the maximum and minimum. This provides padding between the axes and the data.

Figure 21.41 Multiple Panels Using PROC SGRENDER

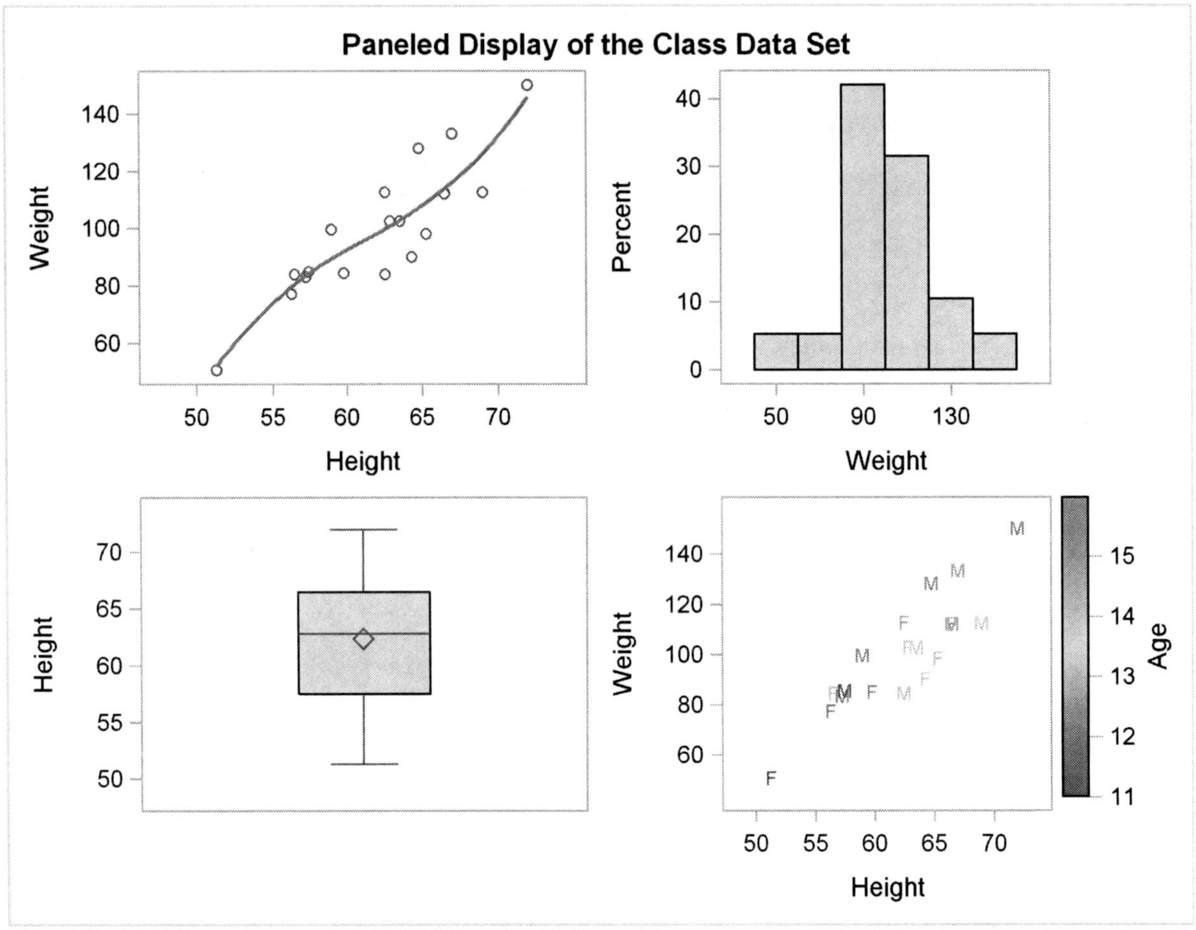

Many other types of graphs are available with the SG procedures. However, even the few examples provided here show the power and flexibility available for making professional-quality statistical graphics. See the *SAS/GRAPH: Graph Template Language User's Guide* and the *SAS/GRAPH: Statistical Graphics Procedures Guide* for more information.

Examples of ODS Statistical Graphics

Example 21.1: Creating Graphs with Tool Tips in HTML

This example demonstrates how to request graphics in HTML that are enhanced with tooltip displays, which appear when you move a mouse over certain features of the graph. When you specify the HTML destination and the IMAGEMAP=ON option in the ODS GRAPHICS statement, an image map of coordinates for tooltips is generated along with the HTML output file. Individual graphs are saved as PNG files.

Example 56.2 and Example 56.8 of Chapter 56, "The MIXED Procedure," analyze a data set with repeated growth measurements for 27 children. The following step creates the data set:

```
data pr;
   input Person Gender $ y1 y2 y3 y4 @@;
   y=y1; Age=8;  output;
   y=y2; Age=10; output;
   y=y3; Age=12; output;
   y=y4; Age=14; output;
   drop y1-y4;
   datalines;
 1 F  21.0  20.0  21.5  23.0     2 F  21.0  21.5  24.0  25.5

 ... more lines ...

;
```

The following statements fit a mixed model with random intercepts and slopes for each child:

```
ods graphics on / imagemap=on;
ods html body='b.html' style=statistical;

proc mixed data=pr method=ml plots=boxplot;
   ods select 'Conditional Residuals by Gender';
   class Person Gender;
   model y = Gender Age Gender*Age;
   random intercept Age / type=un subject=Person;
run;

ods html close;
```

The PLOTS=BOXPLOT option in the PROC MIXED statement requests box plots of observed values and residuals for each classification main effect in the model (Gender and Person). Only the by-gender box plots are actually created due to the ODS SELECT statement, which uses the plot label to select the plot. Output 21.1.1 displays the results. Moving the mouse over a box plot displays a tooltip with summary statistics for the class level. Graphics with tooltips are supported for only the HTML destination.

Output 21.1.1 Box Plot with Tool Tips

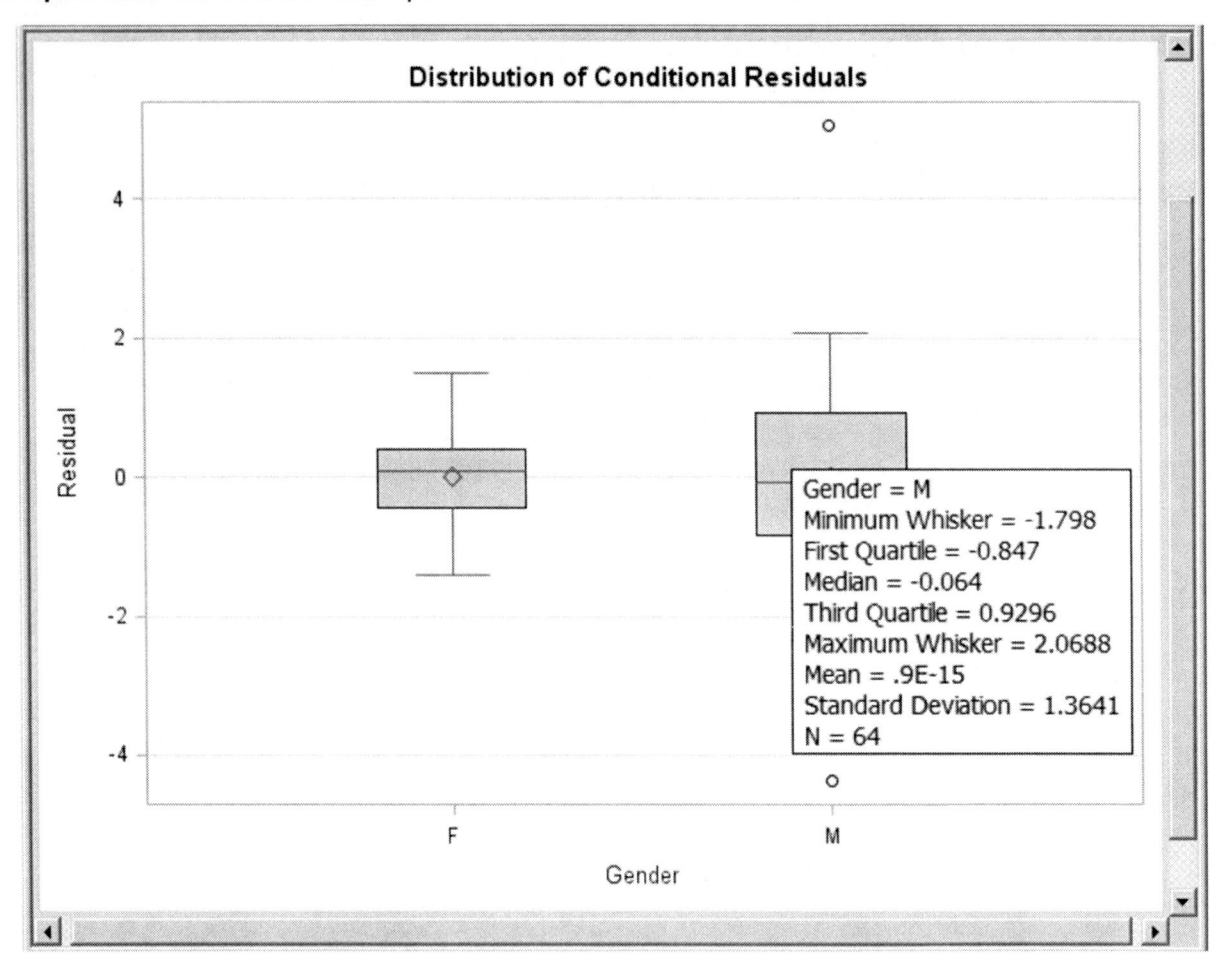

Distribution of Conditional Residuals

Tool tip contents:
- Gender = M
- Minimum Whisker = -1.798
- First Quartile = -0.847
- Median = -0.064
- Third Quartile = 0.9296
- Maximum Whisker = 2.0688
- Mean = .9E-15
- Standard Deviation = 1.3641
- N = 64

Example 21.2: Creating Graphs for a Presentation

The RTF destination provides the easiest way to create ODS graphs for inclusion into a paper or presentation. You can specify the ODS RTF statement to create a file that is easily imported into a document (such as Microsoft Word or WordPerfect) or a presentation (such as Microsoft PowerPoint).

The following statements simulate 100 observations from the model $y = \log(x) + e$, where $x = 1, \ldots, 100$ and e has a normal distribution with mean 0 and variance 1:

```
data one;
   do x = 1 to 100;
      y = log(x) + normal(104);
      output;
   end;
run;
```

The following statements request a loess fit and save the output in the file *loess.rtf*:

```
ods listing close;
ods rtf file="loess.rtf";
ods graphics on;

proc loess data=one;
   model y = x / clm residual;
run;

ods rtf close;
ods listing;
```

The output file includes various tables and the following plots: a plot of the selection criterion versus smoothing parameter, a fit plot with 95% confidence bands, a plot of residual by regressors, and a diagnostics panel. The fit plot is produced with the RTF style and is shown in Output 21.2.1.

Output 21.2.1 Loess Fit Plot with the RTF Style

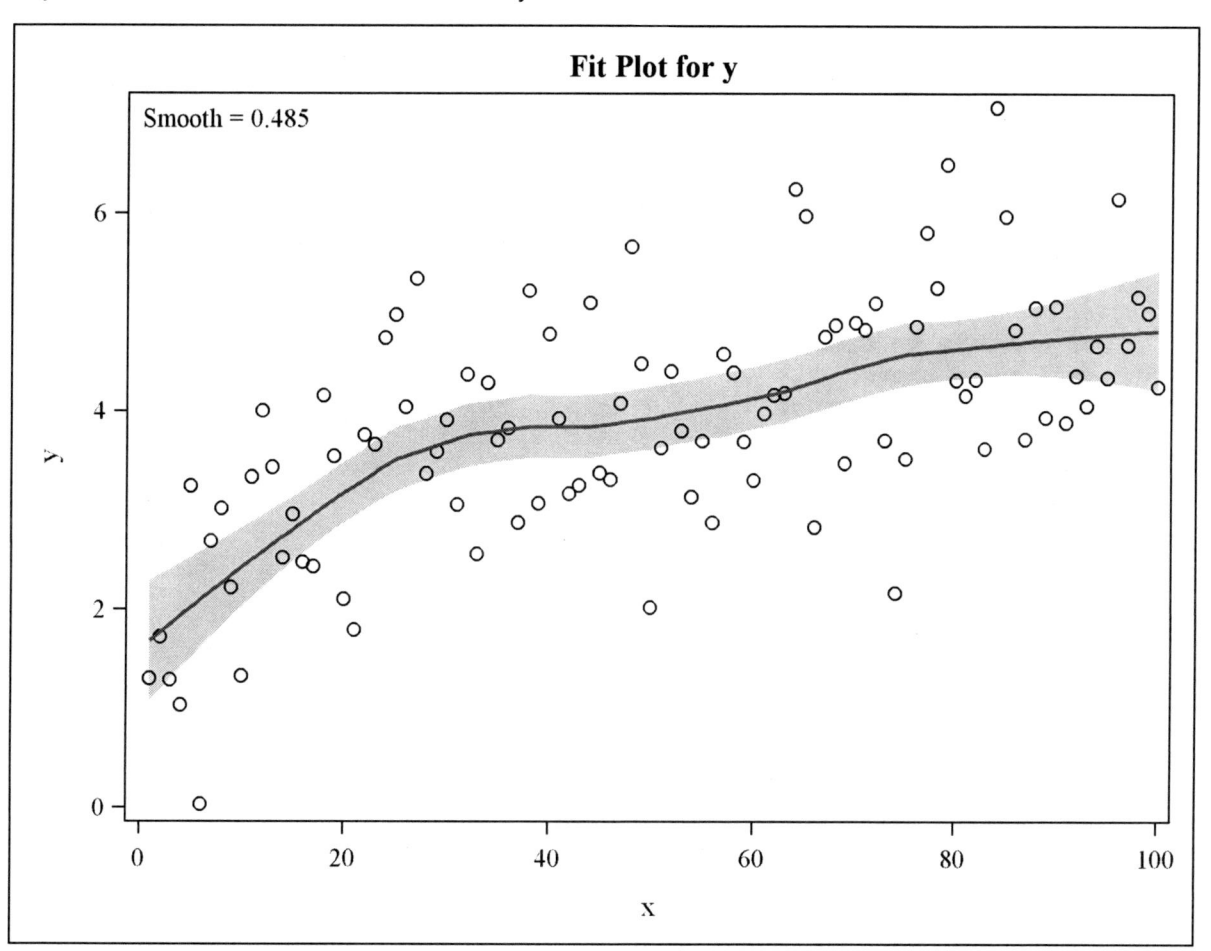

If you are running SAS in the Windows operating system, you can open the RTF file in Microsoft Word and simply copy and paste the graphs into Microsoft PowerPoint. In general, RTF output is convenient for exchange of graphical results between Windows applications through the clipboard.

Alternatively, if you request ODS Graphics with the LISTING or HTML destinations, then your individual graphs are created as PNG files by default. You can insert these files into a Microsoft PowerPoint presentation. See the sections "Naming Graphics Image Files" on page 652 and "Saving Graphics Image Files" on page 654 for information about how the image files are named and saved.

Example 21.3: Creating Graphs in PostScript Files

This example illustrates how to create individual graphs in PostScript files. This is particularly useful when you want to include them in a LaTeX document. Consider again the data set **Stack** created by the following statements:

```
data stack;
   input   x1 x2 x3 y @@;
   datalines;
80  27  89  42     80  27  88  37      75  25  90  37

   ... more lines ...

;
```

The following statements close the LISTING destination, open the LATEX destination with the JOURNAL style, and request a histogram of standardized robust residuals computed with PROC ROBUSTREG:

```
ods graphics on / reset=index;
ods listing close;
ods latex style=Journal;

proc robustreg data=stack plots=histogram;
   model y = x1 x2 x3;
run;

ods latex close;
ods listing;
```

The JOURNAL style displays gray-scale graphs that are suitable for a journal. When you specify the ODS LATEX destination, ODS creates a PostScript file for each individual graph in addition to a LaTeX source file that includes the tabular output and references to the PostScript files. By default, these files are saved in the SAS current folder. The histogram shown in Output 21.3.1 is saved by default in a file named *Histogram.ps*. See the section "Naming Graphics Image Files" on page 652 for details about how graphics image files are named. If both the LISTING destination (which is open unless you explicitly close it) and the LATEX destination are open, then two files are created: *Histogram.ps* and *Histogram1.ps*. If the RESET=INDEX option is not specified in the ODS GRAPHICS statement, and if you run the step more than once, the final name is based on an incremented index.

Output 21.3.1 Histogram Using the JOURNAL Style

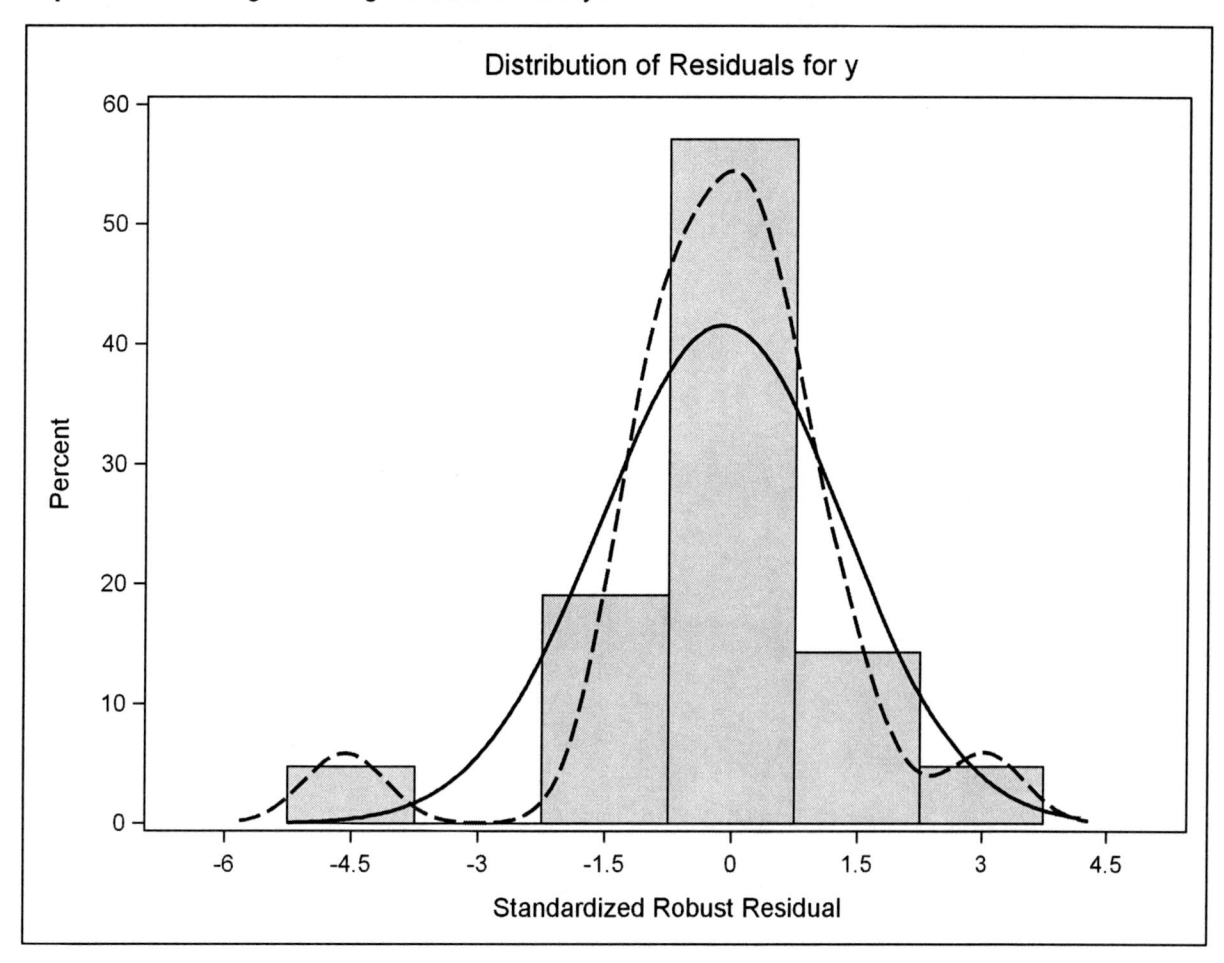

Distribution of Residuals for y

You can use the JOURNAL2 style for a different appearance—the bars are not filled. The following step produces Output 21.3.2:

```
ods graphics on / reset=index;
ods listing close;
ods latex style=Journal2;

proc robustreg data=stack plots=histogram;
   model y = x1 x2 x3;
run;

ods latex close;
ods listing;
```

Output 21.3.2 Histogram Using the JOURNAL2 Style

Distribution of Residuals for y

If you are writing a paper, you can include the graphs in your own LaTeX source file by referencing the names of the individual PostScript graphics files. In this situation, you might not find it necessary to use the LaTeX source file created by SAS. Alternatively, you can include PNG files into a LaTeX document, after using some other ODS destination (such as HTML) to create the PNG files.

Example 21.4: Displaying Graphs Using the DOCUMENT Procedure

This example illustrates the use of the ODS DOCUMENT destination and the DOCUMENT procedure to display your ODS graphs. You can use this approach whenever you want to generate and save your output (both tables and graphs) and then display or replay it later, potentially in subsets or more than once. This approach is particularly useful when you want to display your output in multiple ODS destinations, or when you want to use different styles without rerunning your SAS program. This approach is also useful when you want to break your output into separate parts for inclusion into different parts of a document such as a LaTeX file.

Consider again the data set Stack created by the following statements:

```
data stack;
   input  x1 x2 x3 y @@;
   datalines;
80  27  89  42     80  27  88  37     75  25  90  37

   ... more lines ...

;
```

The following statements request a Q-Q plot from PROC ROBUSTREG with the Stack data:

```
ods listing close;
ods document name=QQDoc(write);

proc robustreg data=stack plots=qqplot;
   model y = x1 x2 x3;
run; quit;

ods document close;
ods listing;
```

The ODS DOCUMENT statement opens an ODS document named QQDoc. All of the results—tables, graphs, titles, notes, footnotes, headers—are stored in the ODS document. None of them are displayed since no other destination is open. In order to display the Q-Q plot with PROC DOCUMENT, you first need to determine its name. You can do this by specifying the ODS TRACE ON statement prior to the procedure statements (see the section "Determining Graph Names and Labels" on page 647 for more information). Alternatively, you can type **odsdocuments** (or **odsd** for short) in the command line to open the Documents window, which you can then use to manage your ODS documents.

The following statements specify an HTML destination and display the residual Q-Q plot by using the REPLAY statement in PROC DOCUMENT:

```
ods html body='b.htm';

proc document name=QQDoc;
   ods select QQPlot;
   replay;
run; quit;

ods html close;
```

Subsequent steps can replay one or more objects from the same ODS document. By default, the REPLAY statement attempts to display every output object stored in the ODS document, but here only the Q-Q plot is displayed because it is specified by the ODS SELECT statement. The plot is displayed in Output 21.4.1.

Output 21.4.1 Q-Q Plot Displayed by PROC DOCUMENT

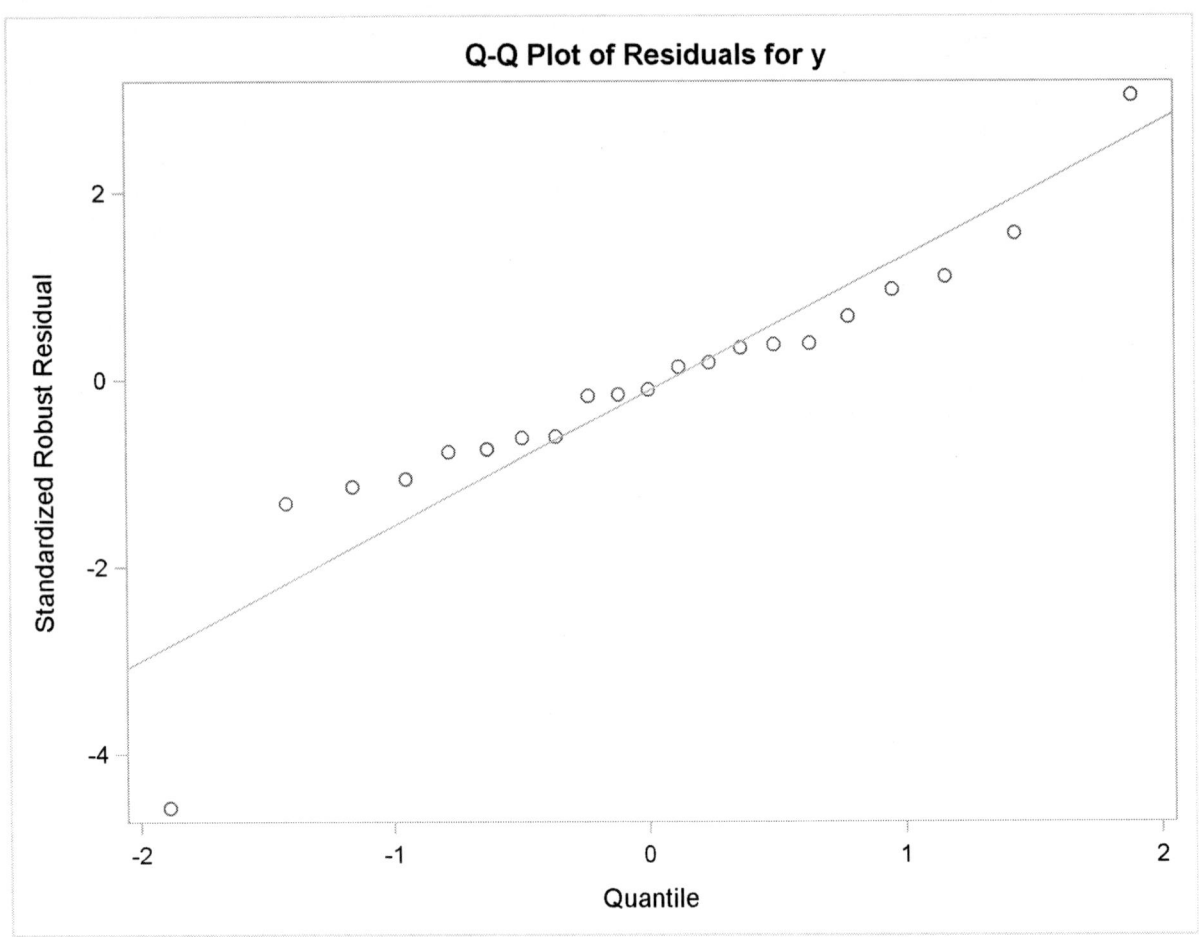

As an alternative to running PROC DOCUMENT with an ODS SELECT statement, you can run PROC DOCUMENT with a *document path* for the Q-Q plot in the REPLAY statement. This approach is preferable when the ODS document contains a large volume of output, so that PROC DOCUMENT does not attempt to process every piece of output stored in the ODS document.

You can determine the ODS document path for the Q-Q plot by specifying the LIST statement with the LEVELS=ALL option in PROC DOCUMENT as follows:

```
proc document name=QQDoc;
   list / levels=all;
run; quit;
```

The contents of the ODS document QQDoc are shown in Output 21.4.2.

Output 21.4.2 Contents of the ODS Document QQDoc

```
Listing of: \Work.Qqdoc\
Order by: Insertion
Number of levels: All

   Obs    Path                                              Type
---------------------------------------------------------------------
     1    \Robustreg#1                                      Dir
     2    \Robustreg#1\ModelInfo#1                          Table
     3    \Robustreg#1\NObs#1                               Table
     4    \Robustreg#1\ParmInfo#1                           Table
     5    \Robustreg#1\SummaryStatistics#1                  Table
     6    \Robustreg#1\ParameterEstimates#1                 Table
     7    \Robustreg#1\DiagSummary#1                        Table
     8    \Robustreg#1\DiagnosticPlots#1                    Dir
     9    \Robustreg#1\DiagnosticPlots#1\QQPlot#1           Graph
    10    \Robustreg#1\GoodFit#1                            Table
```

The ODS document path of the **QQPlot** entry in the **QQDoc** ODS document, as shown in Output 21.4.2, is **\Robustreg#1\DiagnosticPlots#1\QQPlot#1**.

You can use this path to display the residual Q-Q plot with PROC DOCUMENT as follows:

```
proc document name=QQDoc;
    replay \Robustreg#1\DiagnosticPlots#1\QQPlot#1;
run; quit;
```

You can also determine the ODS document path from the Results window or the Documents window. Right-click the object icon and select **Properties**.

The SAS/STAT documentation process uses the ODS document. SAS output is saved into an ODS document that is then replayed into sections of the documentation, which is prepared using LaTeX. In general, when you send your output to the DOCUMENT destination, you can use PROC DOCUMENT to rearrange, duplicate, or remove output from the results of a procedure or a database query. For more information, see the ODS DOCUMENT statement in the section "Dictionary of ODS Language Statements" and the chapter "The DOCUMENT Procedure" in the *SAS Output Delivery System: User's Guide*.

Example 21.5: Customizing Graphs Through Template Changes

This example shows how to use PROC TEMPLATE to customize the appearance and content of an ODS graph. It is divided into several parts; each part illustrates a different aspect of the template that you can easily change. You are never required to change a template, but you can if you want to change aspects of the plot.

Modifying Graph Titles and Axis Labels

This section illustrates the discussion in the section "Graph Templates" on page 687 in the context of changing the default title and Y-axis label for a Q-Q plot created with PROC ROBUSTREG. Consider again the data set Stack created by the following statements:

```
data stack;
   input  x1 x2 x3 y @@;
   datalines;
80  27  89  42      80  27  88  37      75  25  90  37

   ... more lines ...

;
```

The following statements request a Q-Q plot for robust residuals created by PROC ROBUSTREG:

```
ods trace on;
ods graphics on;

proc robustreg data=stack plots=qqplot;
   ods select QQPlot;
   model y = x1 x2 x3;
run;

ods trace off;
```

The Q-Q plot is shown in Output 21.5.1.

Output 21.5.1 Default Q-Q Plot from PROC ROBUSTREG

The ODS TRACE ON statement requests a record of all the ODS output objects created by PROC ROBUSTREG. The trace output is as follows:

```
Output Added:
-------------
Name:       QQPlot
Label:      Residual Q-Q Plot
Template:   Stat.Robustreg.Graphics.QQPlot
Path:       Robustreg.DiagnosticPlots.QQPlot
-------------
```

ODS Graphics creates the Q-Q plot from an ODS data object named **QQPlot** and a graph template named **Stat.Robustreg.Graphics.QQPlot**, which is the default template provided by SAS. Default templates supplied by SAS are saved in the SASHELP.Tmplmst template store (see the section "Graph Templates" on page 687).

To display the default template definition, open the Templates window by typing **odstemplates** (or **odst** for short) in the command line. Expand SASHELP.Tmplmst and click the **Stat** folder. Output 21.5.2 shows the contents of the **Stat** folder.

Output 21.5.2 The Template Window

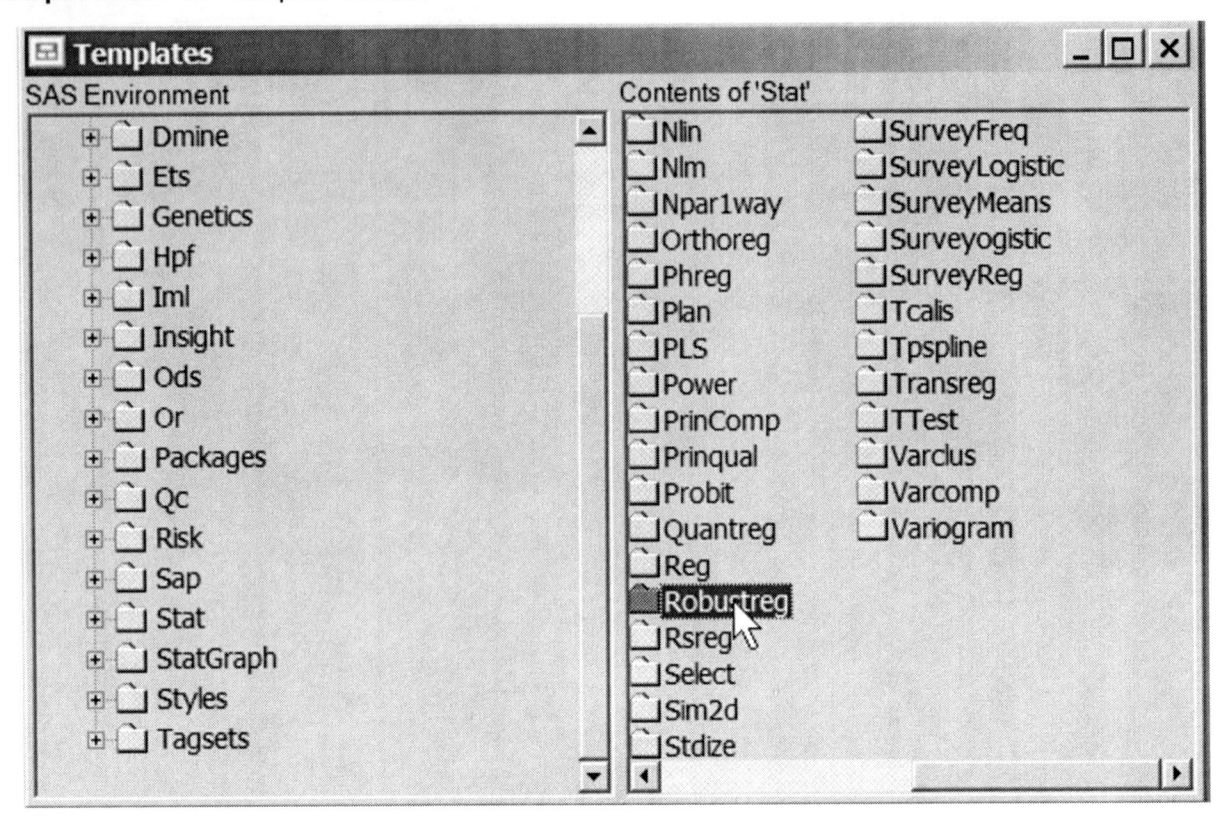

![Template Window screenshot showing SAS Environment tree on the left with folders Dmine, Ets, Genetics, Hpf, Iml, Insight, Ods, Or, Packages, Qc, Risk, Sap, Stat, StatGraph, Styles, Tagsets. The right panel shows Contents of 'Stat': Nlin, Nlm, Npar1way, Orthoreg, Phreg, Plan, PLS, Power, PrinComp, Prinqual, Probit, Quantreg, Reg, Robustreg (selected), Rsreg, Select, Sim2d, Stdize, SurveyFreq, SurveyLogistic, SurveyMeans, Surveyogistic, SurveyReg, Tcalis, Tpspline, Transreg, TTest, Varclus, Varcomp, Variogram.]

Next, open the **Robustreg** folder and then open the **Graphics** folder. Then right-click the `QQPlot` template icon and select **Open**. This opens the Template Browser window shown in Output 21.5.3. You can copy this template to an editor to edit it.

Output 21.5.3 Default Template Definition for Q-Q Plot

```
proc template;
   define statgraph Stat.Robustreg.Graphics.QQPlot;
      notes "Q-Q Plot for Standardized Robust Residuals";
      dynamic _DEPLABEL Residual;
      BeginGraph;
         ENTRYTITLE "Q-Q Plot of Residuals for " _DEPLABEL;
         layout Overlay / yaxisopts=(label="Standardized Robust Residual")
               xaxisopts=(label="Quantile");
            SCATTERPLOT y=eval (SORT(DROPMISSING(RESIDUAL))) x=eval (
               PROBIT((NUMERATE(SORT(DROPMISSING(RESIDUAL))) -0.375)/(0.25 + N
(RESIDUAL)))) / primary=true markerattrs=GRAPHDATADEFAULT rolename=(q=eval (
               PROBIT((NUMERATE(SORT(DROPMISSING(RESIDUAL))) -0.375)/(0.25 + N
(RESIDUAL)))) s=eval (SORT(DROPMISSING(RESIDUAL)))) tip=(q s) tiplabel=(q=
               "Quantile" s="Residual");
            lineparm slope=eval (STDDEV(RESIDUAL)) Y=eval (MEAN(RESIDUAL)) X=0
               / lineattrs=GRAPHREFERENCE extend=true;
         EndLayout;
      EndGraph;
   end;
run;
```

Alternatively, you can submit the following statements to display the `QQPlot` template definition in the SAS log:

```
proc template;
   source Stat.Robustreg.Graphics.QQPlot;
run;
```

The SOURCE statement specifies the fully qualified template name. You can copy and paste the template source into the Program Editor and modify it. The template, with a PROC TEMPLATE and RUN statement added, is shown next:

```
proc template;
   define statgraph Stat.Robustreg.Graphics.QQPlot;
      notes "Q-Q Plot for Standardized Robust Residuals";
      dynamic _DEPLABEL Residual;
      BeginGraph;
         ENTRYTITLE "Q-Q Plot of Residuals for " _DEPLABEL;
         layout Overlay / yaxisopts=(label="Standardized Robust Residual")
            xaxisopts=(label="Quantile");
            SCATTERPLOT y=eval (SORT(DROPMISSING(RESIDUAL))) x=eval (
               PROBIT((NUMERATE(SORT(DROPMISSING(RESIDUAL))) -0.375)/(0.25
               + N(RESIDUAL)))) / primary=true markerattrs=GRAPHDATADEFAULT
               rolename=(q=eval (
               PROBIT((NUMERATE(SORT(DROPMISSING(RESIDUAL))) -0.375)/(0.25
               + N(RESIDUAL)))) s=eval (SORT(DROPMISSING(RESIDUAL))))
               tip=(q s) tiplabel=(q="Quantile" s="Residual");
            lineparm slope=eval (STDDEV(RESIDUAL)) Y=eval (MEAN(RESIDUAL))
               X=0 / lineattrs=GRAPHREFERENCE extend=true;
         EndLayout;
      EndGraph;
   end;
run;
```

In the template, the default title of the Q-Q plot is specified by the ENTRYTITLE statement. The variable _DepLabel is a dynamic variable that provides the name of the dependent variable in the regression analysis. In this case, the name is y. In this template, the label for the axes are specified by the LABEL= suboption of the YAXISOPTS= option for the LAYOUT OVERLAY statement. In other templates, the axis labels come from the column labels of the X-axis and Y-axis columns of the data object. You can see these labels by specifying ODS OUTPUT with the plot data object and running PROC CONTENTS with the resulting SAS data set.

Suppose you want to change the default title to "Analysis of Residuals", and you want the Y-axis label to display the name of the dependent variable. First, replace the ENTRYTITLE statement with the following statement:

```
entrytitle "Analysis of Residuals";
```

Next, replace the LABEL= suboption with the following:

```
label=("Standardized Robust Residual for " _DEPLABEL)
```

You can use dynamic text variables such as _DepLabel in any text element.

732 ♦ *Chapter 21: Statistical Graphics Using ODS*

You can then submit the modified template definition as you would any SAS program, for example, by selecting **Submit** from the **Run** menu. After submitting the PROC TEMPLATE statements, you should see the following message in the SAS log:

```
NOTE: STATGRAPH 'Stat.Robustreg.Graphics.QQPlot' has been
   saved to: SASUSER.TEMPLAT
```

For more information about graph definitions and the graph template language, see the section "Graph Templates" on page 687.

Finally, resubmit the PROC ROBUSTREG statements to display the Q-Q plot created with your modified template. The following statements create Output 21.5.4:

```
proc template;
   define statgraph Stat.Robustreg.Graphics.QQPlot;
      notes "Q-Q Plot for Standardized Robust Residuals";
      dynamic _DEPLABEL Residual;
      BeginGraph;
         entrytitle "Analysis of Residuals";
         layout Overlay /
            yaxisopts=(label=("Standardized Robust Residual for " _DEPLABEL))
            xaxisopts=(label="Quantile");
            SCATTERPLOT y=eval (SORT(DROPMISSING(RESIDUAL))) x=eval (
               PROBIT((NUMERATE(SORT(DROPMISSING(RESIDUAL))) -0.375)/(0.25
               + N(RESIDUAL)))) / primary=true markerattrs=GRAPHDATADEFAULT
               rolename=(q=eval (
               PROBIT((NUMERATE(SORT(DROPMISSING(RESIDUAL))) -0.375)/(0.25
               + N(RESIDUAL)))) s=eval (SORT(DROPMISSING(RESIDUAL))))
               tip=(q s) tiplabel=(q="Quantile" s="Residual");
            lineparm slope=eval (STDDEV(RESIDUAL)) Y=eval (MEAN(RESIDUAL))
               X=0 / lineattrs=GRAPHREFERENCE extend=true;
         EndLayout;
      EndGraph;
   end;
run;

proc robustreg data=stack plots=qqplot;
   ods select QQPlot;
   model y = x1 x2 x3;
run;
```

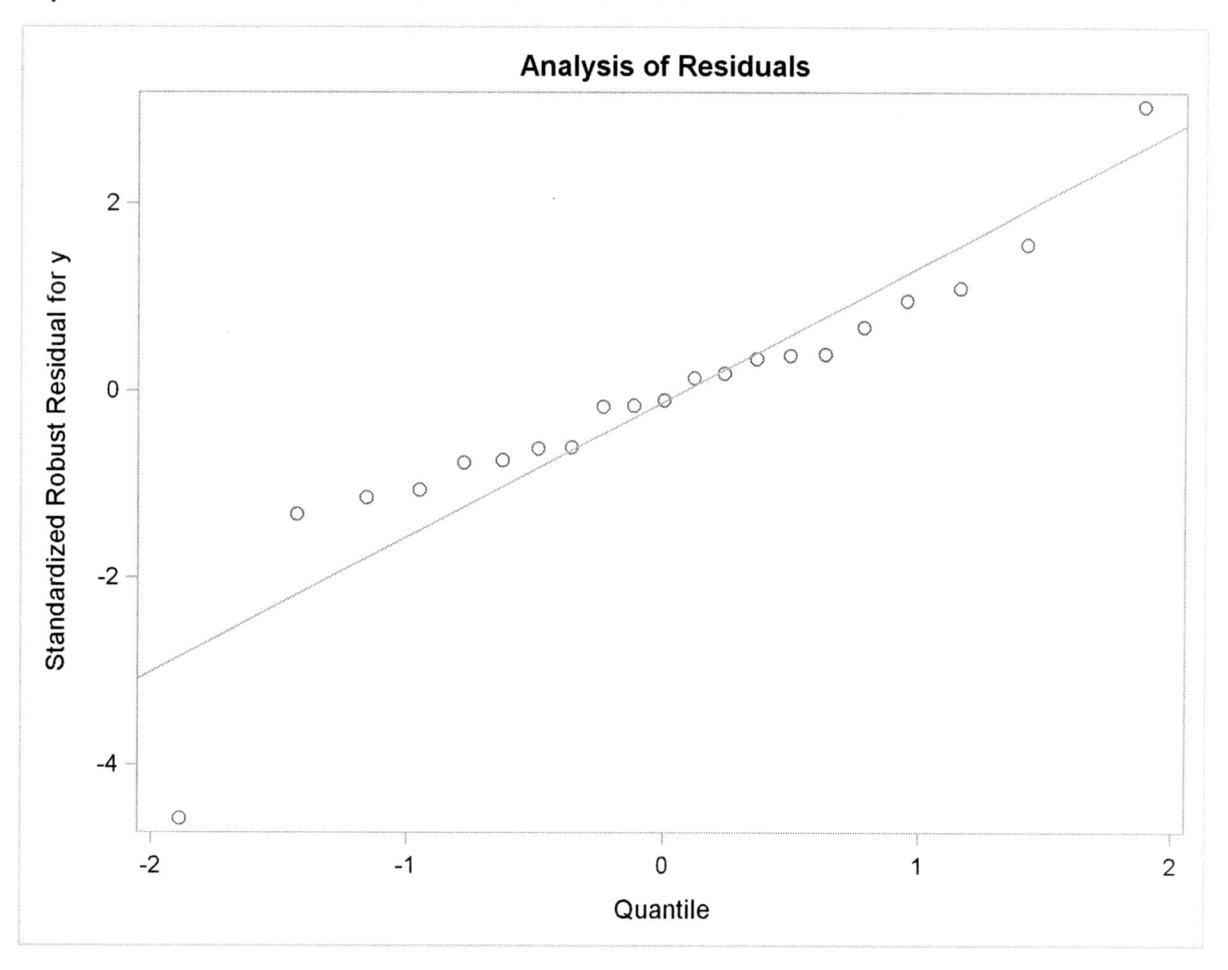

Output 21.5.4 Q-Q Plot with Modified Title and Y-Axis Label

If you have not changed the default template search path, the modified template `QQPlot` is used automatically because SASUSER.Templat occurs before SASHELP.Tmplmst in the ODS search path. See the sections "Saving Customized Templates" on page 697, "Using Customized Templates" on page 697, and "Reverting to the Default Templates" on page 698 for more information about the template search path and the ODS PATH statement.

You do not need to rerun the PROC ROBUSTREG analysis after you modify a graph template if you have stored the plot in an ODS document. After you modify your template, you can submit the PROC DOCUMENT statements in Example 21.4 to replay the Q-Q plot with the modified template. You can run the following statements to revert to the default template:

```
proc template;
   delete Stat.Robustreg.Graphics.QQPlot;
run;
```

Modifying Colors, Line Styles, and Markers

This section shows you how to customize colors, line attributes, and marker symbol attributes by modifying a graph template. In the `QQPlot` template definition shown in Output 21.5.3, the SCATTERPLOT statement specifies a scatter plot of normal quantiles versus ordered standardized residuals. The attributes of the marker symbol in the scatter plot are specified by: `MarkerAttrs=GraphDataDefault`. This is a reference to the style element `GraphDataDefault`. See the section "Style Elements and Attributes" on page 666 for more information.

The actual value of the marker symbol depends on the style that you are using. In this case, since the STATISTICAL style is used, the marker symbol is a circle. You can specify a filled circle as the marker symbol by overriding the symbol portion of the style specification as follows: `MarkerAttrs=GraphDataDefault(symbol=CircleFilled)`.

The value of the SYMBOL= option can be any valid marker symbol or a reference to a style attribute of the form `style-element:attribute`. It is recommended that you use style attributes because they are chosen to provide consistency and appropriate emphasis based on display principles for statistical graphics. If you specify values directly in a template, you are overriding the style and you run the risk of creating a graph that is inconsistent with the style definition. For more information about the syntax of the Graph Template Language and style elements for graphics, see the *SAS/GRAPH: Graph Template Language Reference* and the *SAS Output Delivery System: User's Guide*.

Similarly, you can change the line color and pattern with the LINEATTRS= option in the LINEPARM statement. The LINEPARM statement displays a straight line specified by slope and intercept parameters. The following option changes the color of the line to red and the line pattern to dashed, by overriding those aspects of the style specification: `LineAttrs=GraphReference(color=red pattern=dash)`. To see the results, submit the modified template definition and the PROC ROBUSTREG statements as follows to create Output 21.5.5:

```
proc template;
   define statgraph Stat.Robustreg.Graphics.QQPlot;
      notes "Q-Q Plot for Standardized Robust Residuals";
      dynamic _DEPLABEL Residual;
      BeginGraph;
         entrytitle "Analysis of Residuals";
         layout Overlay /
            yaxisopts=(label=("Standardized Robust Residual for " _DEPLABEL))
            xaxisopts=(label="Quantile");
            SCATTERPLOT y=eval (SORT(DROPMISSING(RESIDUAL))) x=eval(
               PROBIT((NUMERATE(SORT(DROPMISSING(RESIDUAL))) -0.375)/(0.25
               + N(RESIDUAL)))) / primary=true
               markerattrs=GraphDataDefault(symbol=CircleFilled)
               rolename=(q=eval (
               PROBIT((NUMERATE(SORT(DROPMISSING(RESIDUAL))) -0.375)/(0.25
               + N(RESIDUAL)))) s=eval (SORT(DROPMISSING(RESIDUAL))))
               tip=(q s) tiplabel=(q="Quantile" s="Residual");
            lineparm slope=eval (STDDEV(RESIDUAL)) Y=eval (MEAN(RESIDUAL))
               X=0 / lineattrs=GraphReference(color=red pattern=dash)
                  extend=true;
         EndLayout;
      EndGraph;
   end;
run;
```

Example 21.5: Customizing Graphs Through Template Changes ✦ 735

```
ods graphics on;

proc robustreg data=stack plots=qqplot;
   ods select QQPlot;
   model y = x1 x2 x3;
run;
```

Output 21.5.5 Q-Q Plot with Modified Marker Symbols and Line

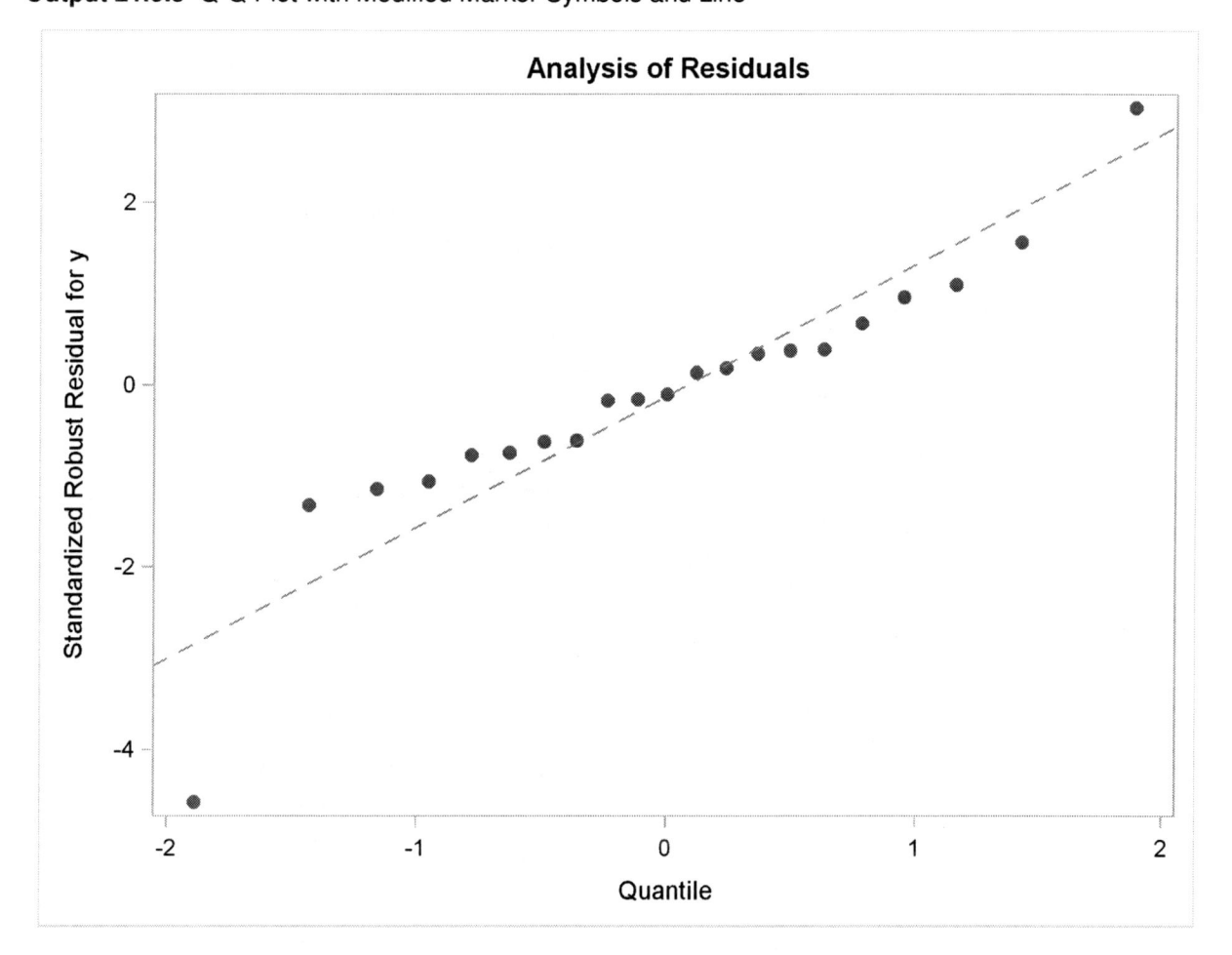

Alternatively, you can replay the plot with PROC DOCUMENT, as in Example 21.4.

Modifying Tick Marks and Grid Lines

This section illustrates how to modify axis tick marks and control grid lines. For example, you can specify the following statement to request tick marks ranging from –4 to 2 in the Y-axis:

```
layout Overlay / yaxisopts=(linearopts=(tickvaluelist=(-4 -3 -2 -1 0 1 2)));
```

The LINEAROPTS= option is used for standard linearly scaled axes (as opposed to log-scaled axes). You use the TICKVALUELIST= to specify the tick marks.

You can control the grid lines by using the GRIDDISPLAY= suboption in the YAXISOPTS= option. Typically, you specify either GRIDDISPLAY=AUTO_OFF (grid lines are not displayed unless the `GraphGridLines` element in the current style contains `DisplayOpts="ON"`) or GRIDDISPLAY=AUTO_ON (grid lines are displayed unless the `GraphGridLines` element in the current style contains `DisplayOpts="OFF"`). Here, the template is modified by specifying GRIDDISPLAY=AUTO_ON for both axes. The following statements produce Output 21.5.6:

```
proc template;
   define statgraph Stat.Robustreg.Graphics.QQPlot;
      notes "Q-Q Plot for Standardized Robust Residuals";
      dynamic _DEPLABEL Residual;
      BeginGraph;
         entrytitle "Analysis of Residuals";
         layout Overlay / yaxisopts=(gridDisplay=Auto_On
            linearopts=(tickvaluelist=(-4 -3 -2 -1 0 1 2))
            label=("Standardized Robust Residual for " _DEPLABEL))
            xaxisopts=(gridDisplay=Auto_On label="Quantile");
            SCATTERPLOT y=eval (SORT(DROPMISSING(RESIDUAL))) x=eval(
               PROBIT((NUMERATE(SORT(DROPMISSING(RESIDUAL))) -0.375)/(0.25
               + N(RESIDUAL)))) / primary=true
               markerattrs=GraphDataDefault(symbol=CircleFilled)
               rolename=(q=eval (
               PROBIT((NUMERATE(SORT(DROPMISSING(RESIDUAL))) -0.375)/(0.25
               + N(RESIDUAL)))) s=eval (SORT(DROPMISSING(RESIDUAL))))
               tip=(q s) tiplabel=(q="Quantile" s="Residual");
            lineparm slope=eval (STDDEV(RESIDUAL)) Y=eval (MEAN(RESIDUAL))
               X=0 / lineattrs=GraphReference(color=red pattern=dash)
                  extend=true;
         EndLayout;
      EndGraph;
   end;
run;

ods graphics on;

proc robustreg data=stack plots=qqplot;
   ods select QQPlot;
   model y = x1 x2 x3;
run;
```

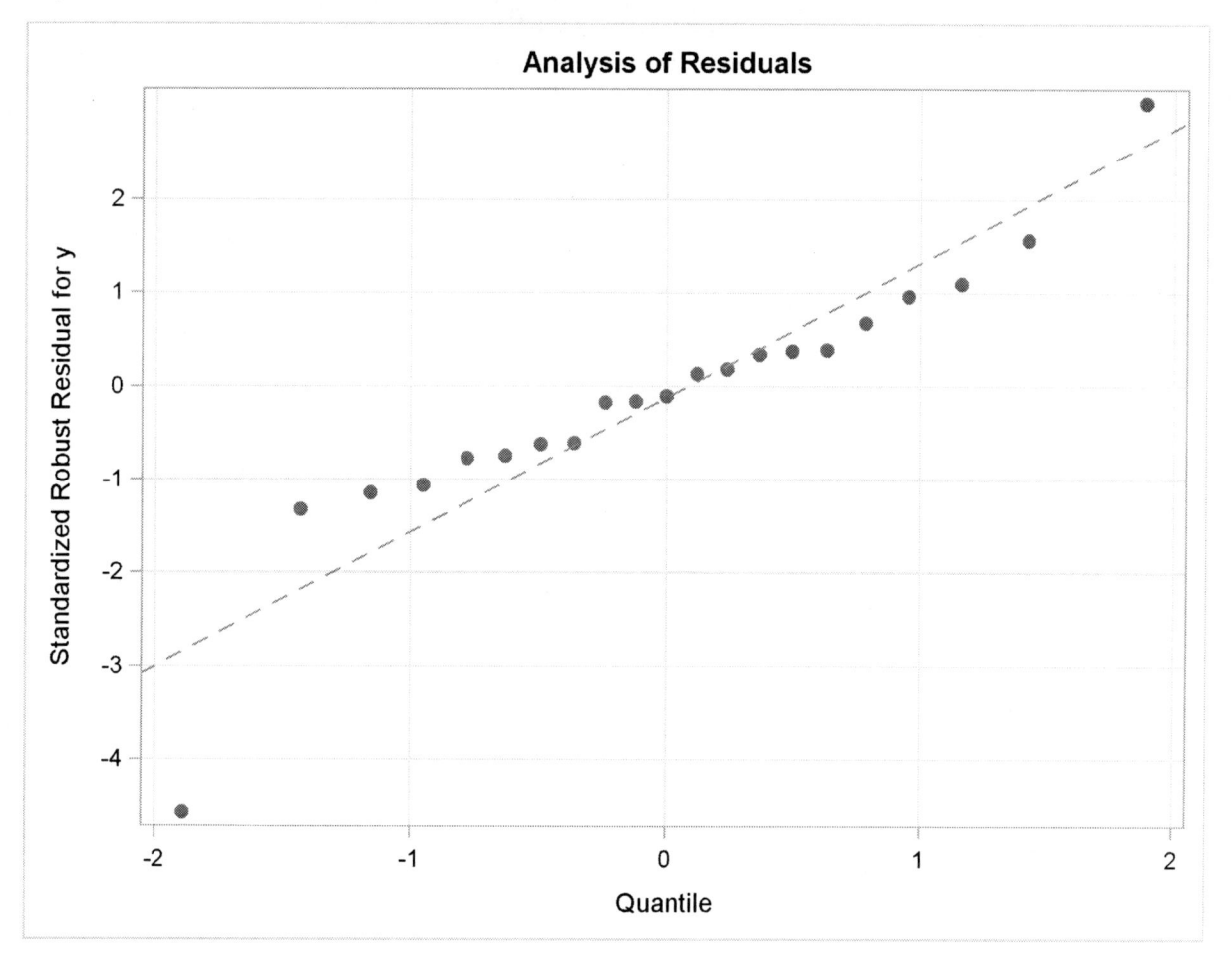

Output 21.5.6 Q-Q Plot with Modified Y-Axis Tick Marks and Grids

You can restore the default template by running the following step:

```
proc template;
   delete Stat.Robustreg.Graphics.QQPlot;
run;
```

See the section "Modifying the Style to Show Grid Lines" on page 737 for more information about grid lines.

Modifying the Style to Show Grid Lines

The section "Modifying Tick Marks and Grid Lines" on page 736 explains that grid lines in graphs are controlled both by template options and by the style. Some graphs never display grid lines because they would interfere with the display. Some graphs always display grid lines because they are a critical part of the display. In both cases, grid control is so important that the template writer is not willing to give control to the style. If you want to change the grid display setting for these graphs, you must edit their templates. Most templates, however, let the style control the grid lines. They either do not display grid lines unless the style forces them on, or they display grid lines unless the style forces them off. The STATISTICAL, DEFAULT, and most other styles use the setting

`DisplayOpts = "Auto"`. Then templates that specify GRIDDISPLAY=AUTO_OFF (the default) do not display grid lines, and templates that specify GRIDDISPLAY=AUTO_ON do display grid lines. You can easily make a new style with `DisplayOpts = "On"` or `DisplayOpts = "Off"` if you would prefer to see grid lines more or less often. This example shows how to set `DisplayOpts = "On"`.

First, you need to find the style source for setting grid lines. The following step displays the STATISTICAL and DEFAULT styles:

```
proc template;
   source Styles.Statistical;
   source Styles.Default;
run;
```

The advantage of displaying both styles together is you can do one search of the results. If grids are defined in the STATISTICAL style, you will find that first. Otherwise, you will first find the definition in the DEFAULT style. An abridged version of the results follows:

```
. . .
class GraphGridLines /
   displayopts = "auto"
   linethickness = 1px
   linestyle = 1
   contrastcolor = GraphColors('ggrid')
   color = GraphColors('ggrid');
. . .
```

You can use this to create a new style that inherits from the STATISTICAL style, but sets the display options for grids to ON, as in the following example:

```
proc template;
   define style Styles.MyGrids;
      parent=styles.statistical;
      class GraphGridLines /
         displayopts = "on"
         linethickness = 1px
         linestyle = 1
         contrastcolor = GraphColors('ggrid')
         color = GraphColors('ggrid');
   end;
run;
```

You can use this new style as in the following example:

```
ods graphics on;
ods listing style=mygrids;

proc robustreg data=stack plots=qqplot;
   ods select QQPlot;
   model y = x1 x2 x3;
run;
```

The preceding statements produce Output 21.5.7, which shows the Q-Q plot with grid lines displayed. The default graph template, supplied by SAS, is used because the custom template created in the section "Modifying Tick Marks and Grid Lines" on page 736 is deleted at the end of that section.

Output 21.5.7 A Style that Makes Grid Lines the Typical Default

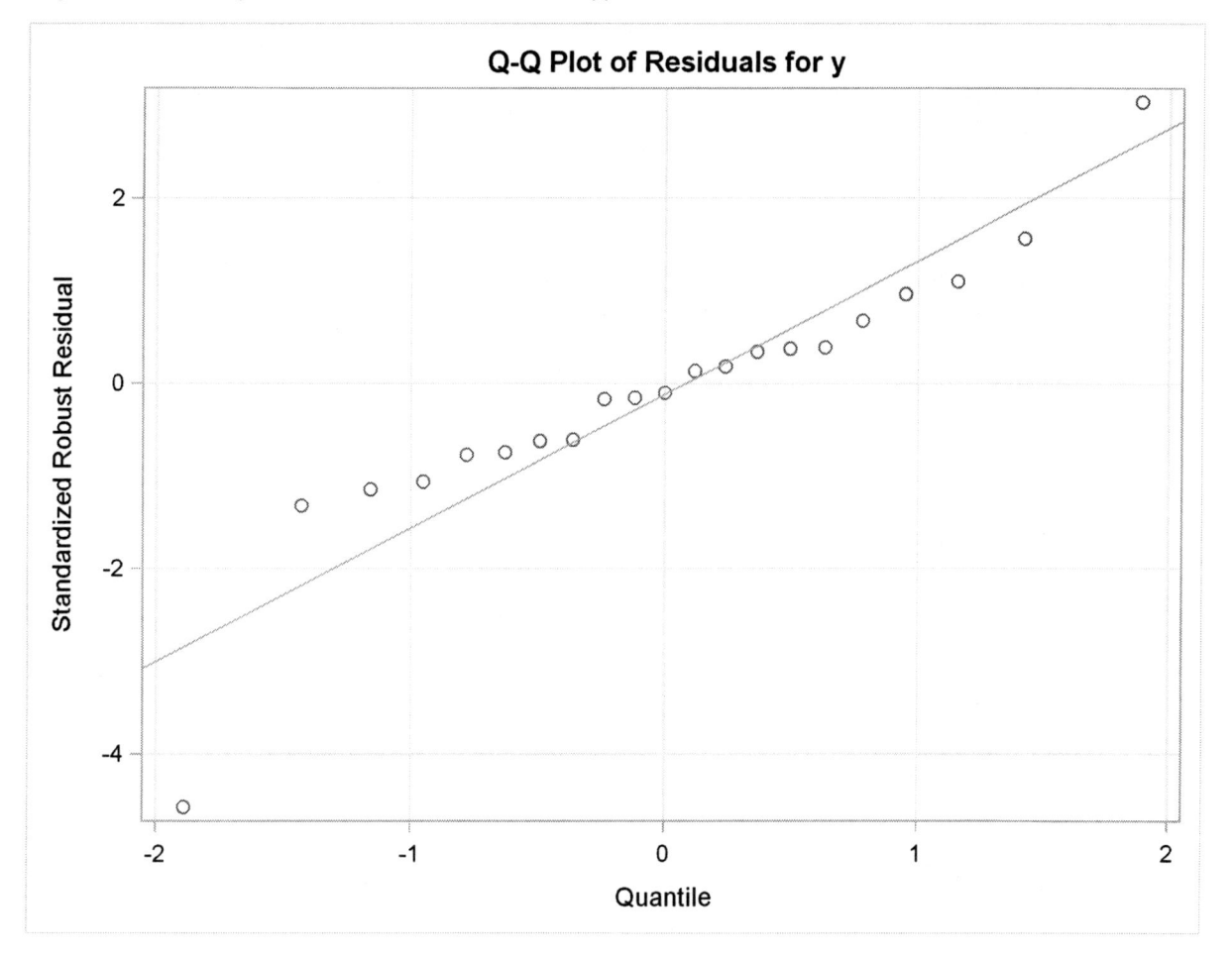

Example 21.6: Customizing Survival Plots

PROC LIFETEST, like other statistical procedures, provides a PLOTS= option and other options for modifying its output without requiring template changes. See Chapter 49, "The LIFETEST Procedure," for more information. Those options are sufficient for most purposes. This example shows ways that you can change the template when those options are not sufficient.

This example changes the default title of the survival plot in PROC LIFETEST from "Product-Limit Survival Estimate" to "Kaplan-Meier Plot" through a template change. Subsequent parts of this example change other aspects of the plot including the legend, inset table, ticks, and the overall layout—all through template changes.

This example uses the BMT data set from the section "Survival Estimate Plot with PROC LIFETEST" on page 612. The following steps create the SAS data set:

```
proc format;
   value risk 1='ALL' 2='AML-Low Risk' 3='AML-High Risk';
run;

data BMT;
   input Group T Status @@;
   format Group risk.;
   label T='Disease Free Time';
   datalines;
1 2081 0 1 1602 0 1 1496 0 1 1462 0 1 1433 0

   ... more lines ...

;
```

The following statements run PROC LIFETEST to determine the template name:

```
ods graphics on;
ods trace output;

proc lifetest data=BMT plots=survival(cb=hw test);
   time T * Status(0);
   strata Group / test=logrank;
run;

ods trace off;
```

The trace output results (not shown) show that the template name for the survival plot is `Stat.Lifetest.Graphics.ProductLimitSurvival`. The following statements display the template:

```
proc template;
   source Stat.Lifetest.Graphics.ProductLimitSurvival;
run;
```

Modifying the Plot Title

The template source listing is lengthy, but to change the title, you only need to find and modify the ENTRYTITLE statements. Many templates have conditional logic (IF statements, for example) and multiple overlays, so it is important that you find all of the relevant ENTRYTITLE statements.

The following abridged version of the statements show how you redefine the template:

```
proc template;
   define statgraph Stat.Lifetest.Graphics.ProductLimitSurvival;
      dynamic . . .;
      BeginGraph;
         if (NSTRATA=1)
            if (EXISTS(STRATUMID))
               entrytitle "Kaplan-Meier Plot for " STRATUMID;
            else
               entrytitle "Kaplan-Meier Plot";
            endif;
            if (PLOTATRISK)
               entrytitle "with Number of Subjects at Risk" / textattrs=
                  GRAPHVALUETEXT;
            endif;
            layout overlay / xaxisopts=(shortlabel=XNAME offsetmin=.05
               linearopts=(viewmax=MAXTIME)) yaxisopts=(label=
               "Survival Probability" shortlabel="Survival" linearopts=(
               viewmin=0 viewmax=1 tickvaluelist=(0 .2 .4 .6 .8 1.0)));
               . . .
            endlayout;
         else
            entrytitle "Kaplan-Meier Plot";
            if (EXISTS(SECONDTITLE))
               entrytitle SECONDTITLE / textattrs=GRAPHVALUETEXT;
            endif;
            layout overlay / xaxisopts=(shortlabel=XNAME offsetmin=.05
               linearopts=(viewmax=MAXTIME)) yaxisopts=(label=
               "Survival Probability" shortlabel="Survival" linearopts=(
               viewmin=0 viewmax=1 tickvaluelist=(0 .2 .4 .6 .8 1.0)));
               . . .
            endlayout;
         endif;
      EndGraph;
   end;
run;

proc lifetest data=BMT plots=survival(cb=hw test);
   time T * Status(0);
   strata Group / test=logrank;
run;
```

The preceding steps create Output 21.6.1.

Output 21.6.1 Kaplan-Meier Plot

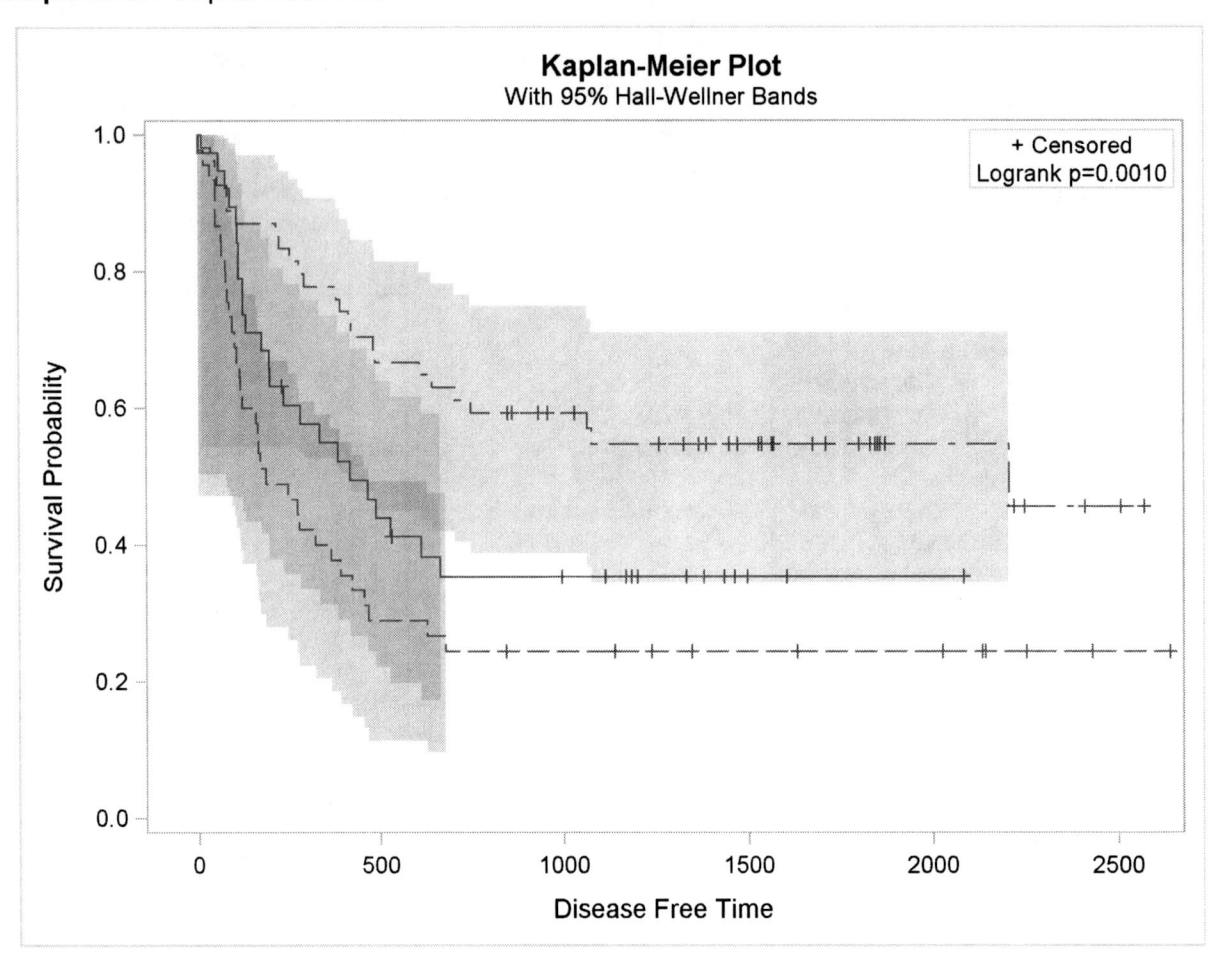

You can restore the default template by running the following step:

```
proc template;
   delete Stat.Lifetest.Graphics.ProductLimitSurvival;
run;
```

Modifying the Axes, Legend, and Inset Table

You can also change other aspects of this plot. The template options `linearopts=(viewmin=0 viewmax=1 tickvaluelist=(0 .2 .4 .6 .8 1.0))` control the minimum value displayed, the maximum value displayed, and the ticks. You can change the range of the vertical axis or the ticks by changing this option everywhere that it occurs. The following specification changes the ticks but not the range of values: `linearopts=(viewmin=0 viewmax=1 tickvaluelist=(0 .25 .5 .75 1.0))`.

There are many other features of this template that you can easily modify. For example, you can change the locations of the inset table and the legend, which are controlled by the following two statements, respectively:

```
layout gridded / rows=2 autoalign=(TOPRIGHT BOTTOMLEFT TOP BOTTOM)
                 border=true BackgroundColor=GraphWalls:Color Opaque=true;

DiscreteLegend "Survival" / title=GROUPNAME location=outside;
```

The LAYOUT GRIDDED statement produces the two-row inset table displayed in the top right corner. The AUTOALIGN= option provides the preferred locations inside the plot for this table, ordered from most preferred to least preferred. You can add new locations or rearrange the existing locations. The DISCRETELEGEND statement places the legend outside of the plot. You can move it inside and print only one legend entry across each row instead of three. This has the effect of changing the orientation of the legend from a row to a column. The modified statements are as follows:

```
layout gridded / rows=2
    autoalign=(BottomRight TOPRIGHT BOTTOMLEFT TOP BOTTOM)
      border=true BackgroundColor=GraphWalls:Color Opaque=true;

DiscreteLegend "Survival" / title=GROUPNAME across=1 location=inside
    autoalign=(TopRight BottomLeft Top Bottom);
```

The full template with these modifications is not shown here, but the new template along with the following statements produce Output 21.6.2:

```
proc lifetest data=BMT plots=survival(cb=hw test);
   time T * Status(0);
   strata Group / test=logrank;
run;
```

Output 21.6.2 Kaplan-Meier Plot with Legend Modifications

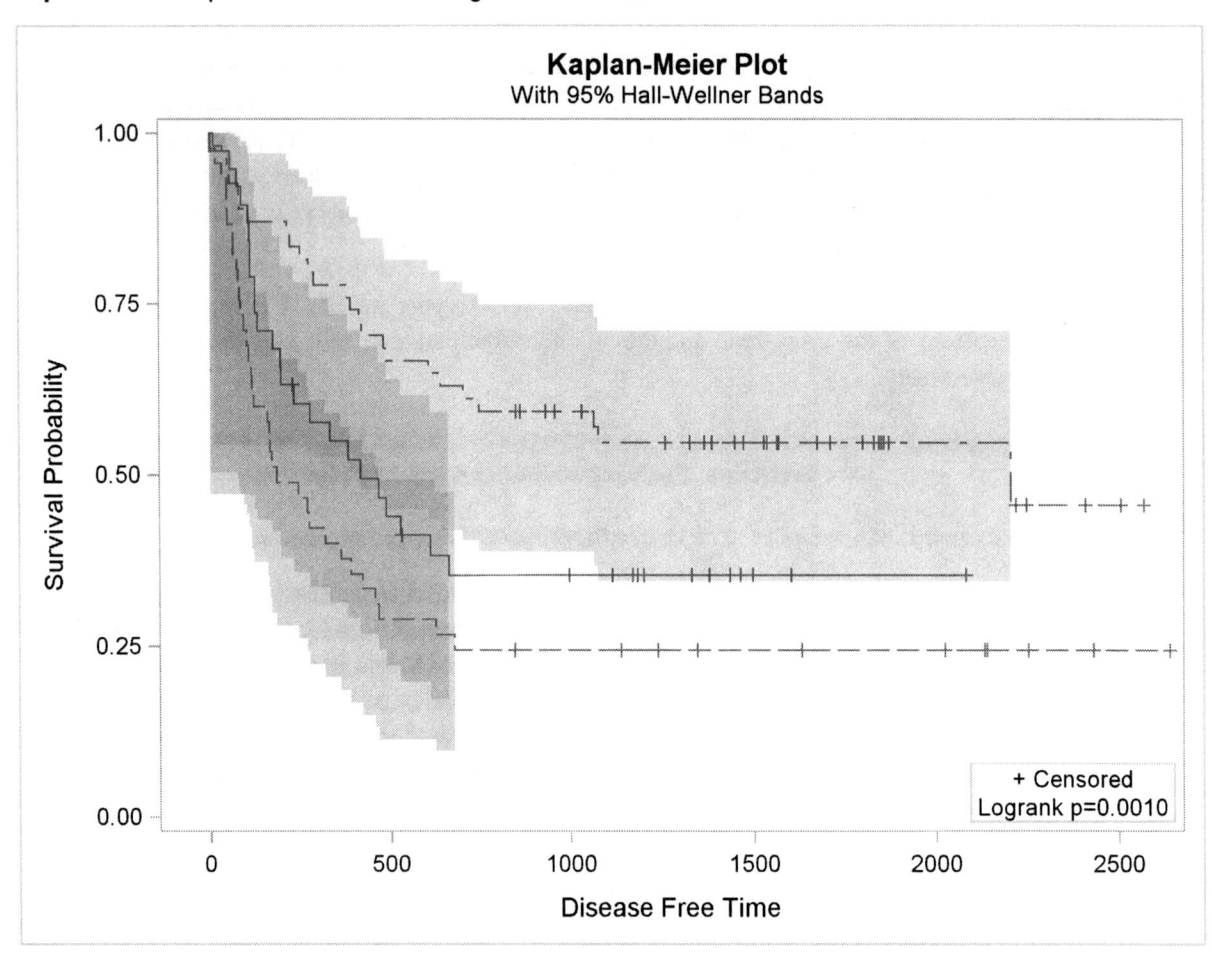

You can restore the default template by running the following step:

```
proc template;
   delete Stat.Lifetest.Graphics.ProductLimitSurvival;
run;
```

Modifying the Layout and Adding a New Inset Table

Example 49.2 of Chapter 49, "The LIFETEST Procedure," uses the following statements to make the plot shown in Output 21.6.3:

```
proc lifetest data=BMT plots=survival(atrisk=0 to 2500 by 500);
   ods select SurvivalPlot;
   time T * Status(0);
   strata Group / test=logrank adjust=sidak;
run;
```

Output 21.6.3 Default Survival Plot with Number of Subjects At Risk

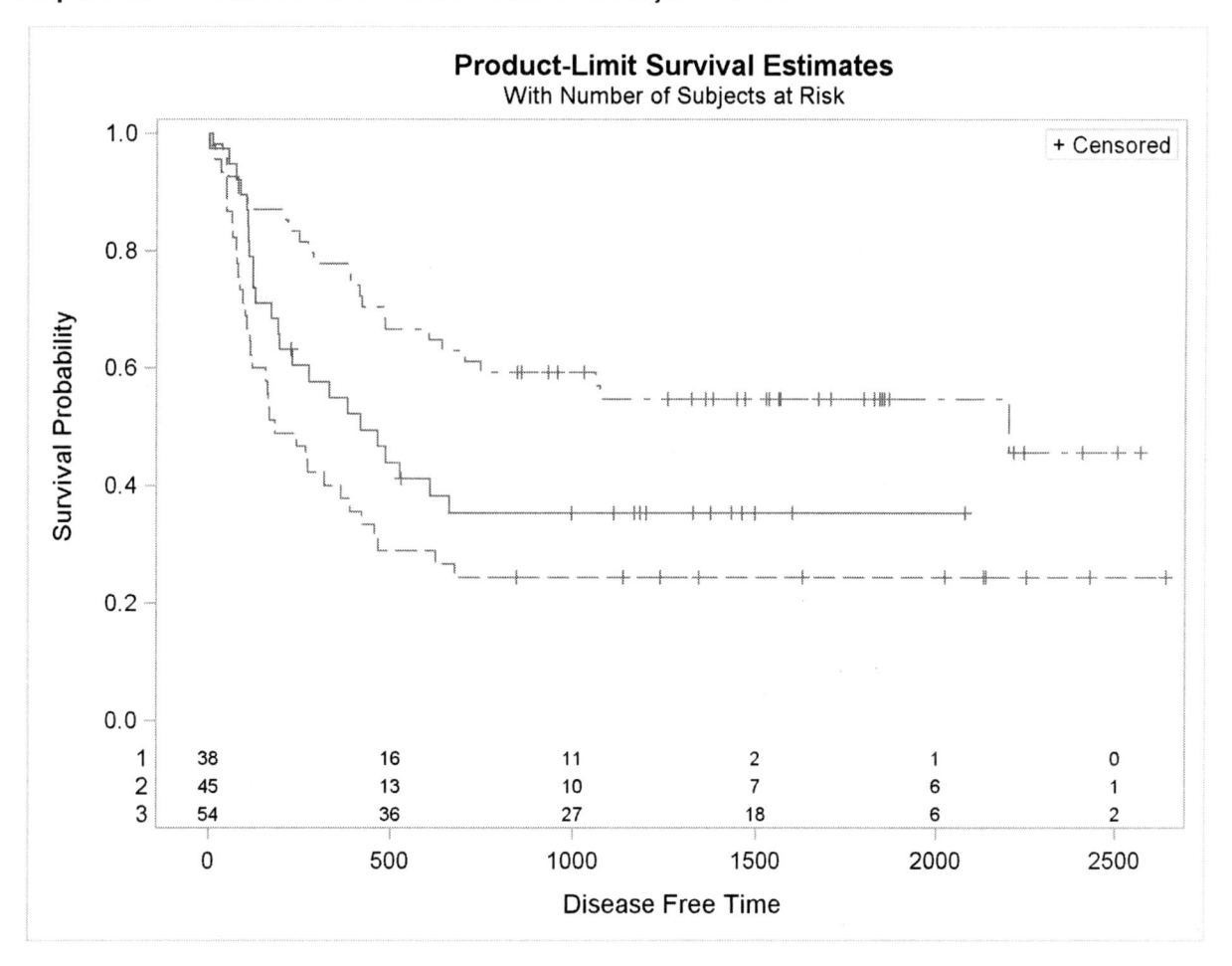

Output 21.6.3 displays the estimated disease-free survival functions for the three leukemia groups with the number of subjects at risk at 0, 500, 1000, 1500, 2000, and 2500 days. The rest of this example shows you how to modify the template to produce the plot displayed in Output 21.6.4. This new plot differs from the old plot in several ways. It has a new inset table in the top right corner with the number of observations and the number of events in the each stratum. The legend has been moved inside the plot and combined with the old inset table that showed the marker for censored observations. The information about the subjects at risk has been moved into a table below the plot. Also, the title change from the first part of the example has been retained.

Output 21.6.4 Kaplan-Meier Plot with a Different Layout

Kaplan-Meier Plot
With Number of Subjects at Risk

	Event	Total
1	24	38
2	34	45
3	25	54

Disease Free Time

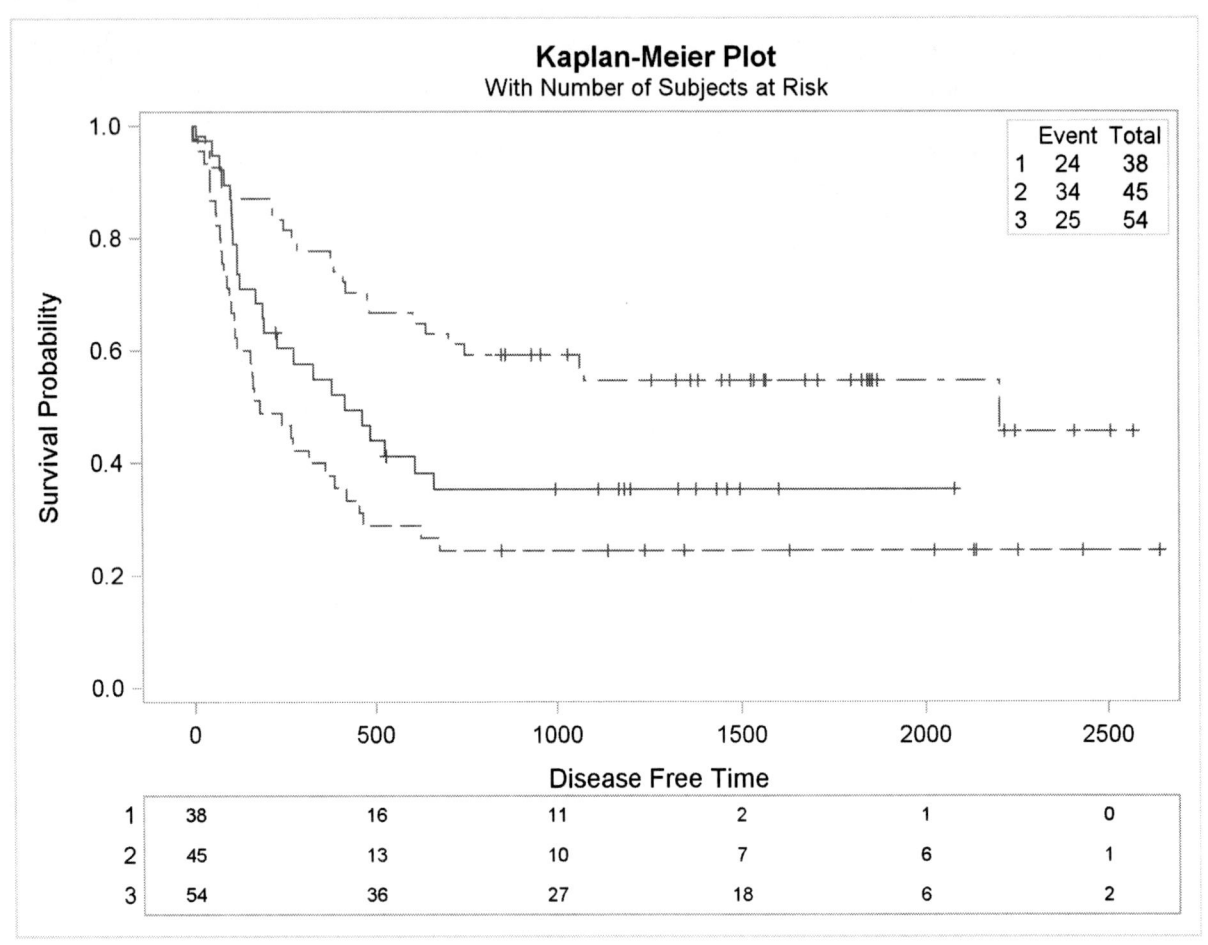

1	38	16	11	2	1	0
2	45	13	10	7	6	1
3	54	36	27	18	6	2

These changes are easy, if they are broken down and performed one step at a time. You can use the template with the new title from the beginning of this example as a starting point for these modifications. Before proceeding, you should notice the outermost layouts of this template, which are shown next:

```
proc template;
   define statgraph Stat.Lifetest.Graphics.ProductLimitSurvival;
       . . .
       BeginGraph;
          if (NSTRATA=1)
             . . .
             layout overlay / . . .;
                . . .
             endlayout;
          else
             . . .
             layout overlay / . . .;
                . . .
             endlayout;
          endif;
       EndGraph;
   end;
run;
```

This template consists of two major parts: a layout that is used when there is only one stratum and a layout that is used with more than one stratum. Every section of this example has more than one stratum, so it is the changes to the second layout (or more precisely the ELSE portion of the template) that are affecting the results.

PROC LIFETEST makes available a series of dynamic variables that it does not display by default. See the section "Additional Dynamic Variables for Survival Plots Using ODS Graphics" on page 3765 in Chapter 49, "The LIFETEST Procedure," for information about these dynamic variables. You can use these dynamic variables to add the new inset table to the plot. The following step creates the table in a gridded layout that is added to the second layout:

```
proc template;
   define statgraph Stat.Lifetest.Graphics.ProductLimitSurvival;
      . . .
      BeginGraph;
         if (NSTRATA=1)
            . . .
            layout overlay / . . .;
               . . .
            endlayout;
         else
            . . .
            layout overlay / . . .;
               . . .
               dynamic NObs1 NObs2 NObs3 NEvent1 NEvent2 NEvent3;
               layout gridded / columns=3 border=TRUE autoalign=(TopRight);
                  entry "";     entry "Event";    entry "Total";
                  entry "1";    entry NEvent1;    entry NObs1;
                  entry "2";    entry NEvent2;    entry NObs2;
                  entry "3";    entry NEvent3;    entry Nobs3;
               endlayout;
            endlayout;
         endif;
      EndGraph;
   end;
run;
```

The at-risk information in Output 21.6.3 is produced by a BLOCKPLOT statement in the second layout. The modified template, with the first layout removed (since it is not needed in this example) and the BLOCKPLOT statement displayed is as follows:

```
proc template;
   define statgraph Stat.Lifetest.Graphics.ProductLimitSurvival;
      . . .
      BeginGraph;
         layout overlay / . . .;
            . . .
            if (PLOTATRISK)
               innermargin / align=bottom;
               blockplot x=TATRISK block=ATRISK / class=CLASSATRISK
                  repeatedvalues=true display=(label values)
                  valuehalign=start valuefitpolicy=truncate
```

```
                   labelposition=left labelattrs=GRAPHVALUETEXT
                   valueattrs=GRAPHDATATEXT(size=7pt)
                   includemissingclass=false;
             endinnermargin;
             endif;
             . . .
             dynamic NObs1 NObs2 NObs3 NEvent1 NEvent2 NEvent3;
             layout gridded / columns=3 border=TRUE autoalign=(TopRight);
                entry "";       entry "Event";    entry "Total";
                entry "1";      entry NEvent1;    entry NObs1;
                entry "2";      entry NEvent2;    entry NObs2;
                entry "3";      entry NEvent3;    entry Nobs3;
             endlayout;
          endlayout;
       EndGraph;
    end;
run;
```

In the next step, the at-risk information is moved out of the plot and into a table below the plot. The template is given a new overall layout—a LAYOUT LATTICE that has two panels stacked vertically, one for the plot and one for the at-risk information. Using ROWWEIGHTS=(.85 .15), the plot on top occupies 85% of the display and the at-risk information in the second panel occupies 15%. The option COLUMNDATARANGE=UNIONALL is used to create a common axis across the two panels. In these next steps, you also move the legend inside (similar to the previous part of this example) and rearrange the three inset boxes. The new template structure is as follows:

```
proc template;
   define statgraph Stat.Lifetest.Graphics.ProductLimitSurvival;
      . . .
      BeginGraph;
         . . .
         layout lattice /  rows=2 columns=1 columndatarange=unionall
                          rowweights=(.85 .15);

            layout overlay / . . .;
               . . .
               DiscreteLegend "Survival" / location=inside
                              autoalign=(BottomRight);
               . . .
               layout gridded / rows=1 autoalign=(BottomLeft)
                     border=true BackgroundColor=GraphWalls:Color
                     Opaque =true;
                  entry "+ Censored";
               endlayout;
               . . .
            dynamic NObs1 NObs2 NObs3 NEvent1 NEvent2 NEvent3;
            layout gridded / columns=3 border=TRUE autoalign=(TopRight);
                entry "";       entry "Event";    entry "Total";
                entry "1";      entry NEvent1;    entry NObs1;
                entry "2";      entry NEvent2;    entry NObs2;
                entry "3";      entry NEvent3;    entry Nobs3;
            endlayout;
         endlayout;
```

```
            layout overlay / xaxisopts=(display=none);
               blockplot x=TATRISK block=ATRISK / class=CLASSATRISK
                     repeatedvalues=true display=(label values)
                     valuehalign=start valuefitpolicy=truncate
                     labelposition=left labelattrs=GRAPHVALUETEXT
                     valueattrs=GRAPHDATATEXT(size=7pt)
                     includemissingclass=false;
            endlayout;
         endlayout;
      EndGraph;
   end;
run;
```

You can further simplify the plot by removing the title from the legend (which is currently the variable name Group) and instead adding "+ Censored" (the contents of the inset table) to the legend in place of the title, as in the following statement:

```
DiscreteLegend "Survival" / title="+ Censored"
   titleattrs=GraphValueText location=inside autoalign=(Bottom);
```

The option TITLEATTRS=GRAPHVALUETEXT is specified so that the "+ Censored" appears in the same font as the other entries in the legend and appears to be just another part of the legend. All of the statements for making the old inset table can now be removed from the template. The full template also plots bands, which are not used in this example, so they can be removed as well. The resulting template is as follows:

```
proc template;
   define statgraph Stat.Lifetest.Graphics.ProductLimitSurvival;

      dynamic NStrata xName plotAtRisk plotCensored plotCL plotHW plotEP
         labelCL labelHW labelEP maxTime StratumID classAtRisk plotBand
         plotTest GroupName yMin Transparency SecondTitle TestName pValue;

   BeginGraph;

      entrytitle "Kaplan-Meier Plot";
      if (EXISTS(SECONDTITLE))
         entrytitle SECONDTITLE / textattrs=GRAPHVALUETEXT;
      endif;

      layout lattice / rows=2 columns=1 columndatarange=unionall
                       rowweights=(.85 .15);

         layout overlay / xaxisopts=(shortlabel=XNAME offsetmin=.05
            linearopts=(viewmax=MAXTIME)) yaxisopts=(label=
            "Survival Probability" shortlabel="Survival" linearopts=(
            viewmin=0 viewmax=1 tickvaluelist=(0 .2 .4 .6 .8 1.0)));

            stepplot y=SURVIVAL x=TIME / group=STRATUM index=STRATUMNUM
               name="Survival" rolename=(_tip1=ATRISK _tip2=EVENT) tip=(y
               x Time _tip1 _tip2);
```

```
            if (PLOTCENSORED)
               scatterplot y=CENSORED x=TIME / group=STRATUM index=
                   STRATUMNUM markerattrs=(symbol=plus);
            endif;

            DiscreteLegend "Survival" / title="+ Censored"
               titleattrs=GraphValueText location=inside
               autoalign=(BOTTOM);

            dynamic NObs1 NObs2 NObs3 NEvent1 NEvent2 NEvent3;
            layout gridded / columns=3 border=TRUE autoalign=(TopRight);
               entry "";         entry "Event";     entry "Total";
               entry "1";        entry NEvent1;     entry NObs1;
               entry "2";        entry NEvent2;     entry NObs2;
               entry "3";        entry NEvent3;     entry Nobs3;
            endlayout;
         endlayout;

         layout overlay / xaxisopts=(display=none);
            blockplot x=TATRISK block=ATRISK / class=CLASSATRISK
                  repeatedvalues=true display=(label values)
                  valuehalign=start valuefitpolicy=truncate
                  labelposition=left labelattrs=GRAPHVALUETEXT
                  valueattrs=GRAPHDATATEXT(size=7pt)
                  includemissingclass=false;
         endlayout;
      endlayout;
   EndGraph;
   end;
run;
```

The following step uses the new template to create the desired plot:

```
proc lifetest data=BMT plots=survival(atrisk=0 to 2500 by 500);
   ods select SurvivalPlot;
   time T * Status(0);
   strata Group / test=logrank adjust=sidak;
   run;
```

The plot is displayed in Output 21.6.4 at the beginning of this section.

This example removed a great deal of functionality from the default template, so that the final, modified template would be relatively simple and understandable. This is not necessary. The template could have been modified without deleting the first LAYOUT OVERLAY and other statements. The strategy for template modification illustrated in this example can be applied to any complicated template: identify the overall structure, isolate the relevant pieces, and then make changes in stages. Since the modified template will no longer work for all analyses, it is important that you delete it when you are done, as in the following example:

```
proc template;
   delete Stat.Lifetest.Graphics.ProductLimitSurvival;
run;
```

Example 21.7: Customizing Panels

This example illustrates how to modify the regression fit diagnostics panel shown in Figure 21.1 so that it displays a subset of the component plots. The original panel consists of eight plots and a summary statistics box. The ODS trace output from PROC REG shown previously shows that the template for the diagnostics panel is `Stat.REG.Graphics.DiagnosticsPanel`. The following statements display the template:

```
proc template;
    source Stat.REG.Graphics.DiagnosticsPanel;
run;
```

An abridged version of the results is shown next:

```
define statgraph Stat.Reg.Graphics.DiagnosticsPanel;
   notes "Diagnostics Panel";
   dynamic . . .;
   BeginGraph / designheight=defaultDesignWidth;
      entrytitle halign=left textattrs=GRAPHVALUETEXT _MODELLABEL
         halign=center textattrs=GRAPHTITLETEXT "Fit Diagnostics"
         " for " _DEPNAME;
      layout lattice / columns=3 rowgutter=10 columngutter=10
         shrinkfonts=true rows=3;
         layout overlay / xaxisopts=(shortlabel='Predicted');
            . . .
         endlayout;
         layout overlay / xaxisopts=(shortlabel='Predicted');
            . . .
         endlayout;
         layout overlay / xaxisopts=(label='Leverage' offsetmax=0.05)
            . . .
         endlayout;
         layout overlay / yaxisopts=(label="Residual" shortlabel=
            "Resid") xaxisopts=(label="Quantile");
            . . .
         endlayout;
         layout overlayequated / xaxisopts=(shortlabel='Predicted')
            . . .
         endlayout;
         layout overlay / xaxisopts=(linearopts=(integer=true) label=
            "Observation" shortlabel="Obs" offsetmax=0.05) yaxisopts=(
            offsetmin=0.05 offsetmax=0.05);
            . . .
         endlayout;
         layout overlay / xaxisopts=(label="Residual") yaxisopts=(label
            ="Percent");
            . . .
         endlayout;
         layout lattice / columns=2 rows=1 rowdatarange=unionall
            columngutter=0;
            . . .
         endlayout;
```

752 ✦ *Chapter 21: Statistical Graphics Using ODS*

```
               if (_SHOWSTATS =1)
                  layout overlay;
                     . . .
               endLayout;
               endif;
               if (_SHOWSTATS = 2)
                  layout overlay / yaxisopts=(gridDisplay=auto_off label=
                     "Residual");
                     . . .
               endlayout;
               endif;
            endlayout;
         EndGraph;
      end;
```

The outermost components of the template are a BEGINGRAPH/ENDGRAPH block with a lattice layout with ROWS=3 and COLUMNS=3 that defines the 3×3 panel of plots. Inside that are nine layouts, one for each cell, the last of which is conditionally defined. The LAYOUT statements define the components of the panel from left to right and top to bottom. You can eliminate some of the panels and produce a 2×2 panel as follows:

```
proc template;
   define statgraph Stat.Reg.Graphics.DiagnosticsPanel;
      notes "Diagnostics Panel";
      dynamic _DEPLABEL _DEPNAME _MODELLABEL _OUTLEVLABEL _TOTFREQ _NPARM
         _NOBS _OUTCOOKSDLABEL _SHOWSTATS _NSTATSCOLS _DATALABEL _SHOWNObs
         _SHOWTOTFREQ _SHOWNParm _SHOWEDF _SHOWMSE _SHOWRSquare
         _SHOWAdjRSq _SHOWSSE _SHOWDepMean _SHOWCV _SHOWAIC _SHOWBIC
         _SHOWCP _SHOWGMSEP _SHOWJP _SHOWPC _SHOWSBC _SHOWSP _EDF _MSE
         _RSquare _AdjRSq _SSE _DepMean _CV _AIC _BIC _CP _GMSEP _JP _PC
         _SBC _SP;
      BeginGraph / designheight=defaultDesignWidth;
         entrytitle halign=left textattrs=GRAPHVALUETEXT _MODELLABEL
            halign=center textattrs=GRAPHTITLETEXT "Fit Diagnostics"
            " for " _DEPNAME;
         layout lattice / columns=2 rowgutter=10 columngutter=10
            shrinkfonts=true rows=2;
            layout overlay / xaxisopts=(shortlabel='Predicted');
               referenceline y=-2;
               referenceline y=2;
               scatterplot y=RSTUDENT x=PREDICTEDVALUE / primary=true
                  datalabel=_OUTLEVLABEL rolename=(_tip1=OBSERVATION _id1=
                  ID1 _id2=ID2 _id3=ID3 _id4=ID4 _id5=ID5) tip=(y x _tip1
                  _id1 _id2 _id3 _id4 _id5);
            endlayout;
            layout overlay / yaxisopts=(label="Residual" shortlabel=
               "Resid") xaxisopts=(label="Quantile");
               lineparm slope=eval (STDDEV(RESIDUAL)) y=eval (
                  MEAN(RESIDUAL)) x=0 / extend=true lineattrs=
                  GRAPHREFERENCE;
               scatterplot y=eval (SORT(DROPMISSING(RESIDUAL))) x=eval (
                  PROBIT((NUMERATE(SORT(DROPMISSING(RESIDUAL))) -0.375)/
                  (0.25 + N(RESIDUAL)))) / markerattrs=GRAPHDATADEFAULT
```

```
                    primary=true
                    rolename=(s=eval (SORT(DROPMISSING(RESIDUAL))) nq=eval (
                    PROBIT((NUMERATE(SORT(DROPMISSING(RESIDUAL))) -0.375)
                    /(0.25 + N(RESIDUAL))))) tiplabel=(nq="Quantile"
                    s="Residual")
                    tip=(nq s);
            endlayout;
            layout overlayequated / xaxisopts=(shortlabel='Predicted')
                    yaxisopts=(label=_DEPLABEL shortlabel="Observed")
                    equatetype=square;
                lineparm slope=1 x=0 y=0 / extend=true lineattrs=
                    GRAPHREFERENCE;
                scatterplot y=DEPVAR x=PREDICTEDVALUE / primary=true
                    datalabel=_OUTLEVLABEL rolename=(_tip1=OBSERVATION _id1=
                    ID1 _id2=ID2 _id3=ID3 _id4=ID4 _id5=ID5) tip=(y x _tip1
                    _id1 _id2 _id3 _id4 _id5);
            endlayout;
            layout overlay / xaxisopts=(label="Residual") yaxisopts=(label
                    ="Percent");
                histogram RESIDUAL / primary=true;
                densityplot RESIDUAL / name="Normal" legendlabel="Normal"
                    lineattrs=GRAPHFIT;
            endlayout;
        endlayout;
      EndGraph;
    end;
run;

proc reg data=sashelp.class;
    model Weight = Height;
run; quit;
```

This template plots the residuals by predicted values, the Q-Q plot, the actual by predicted plot, and the residual histogram. The results are shown in Output 21.7.1.

Output 21.7.1 Diagnostics Panel with Four Plots

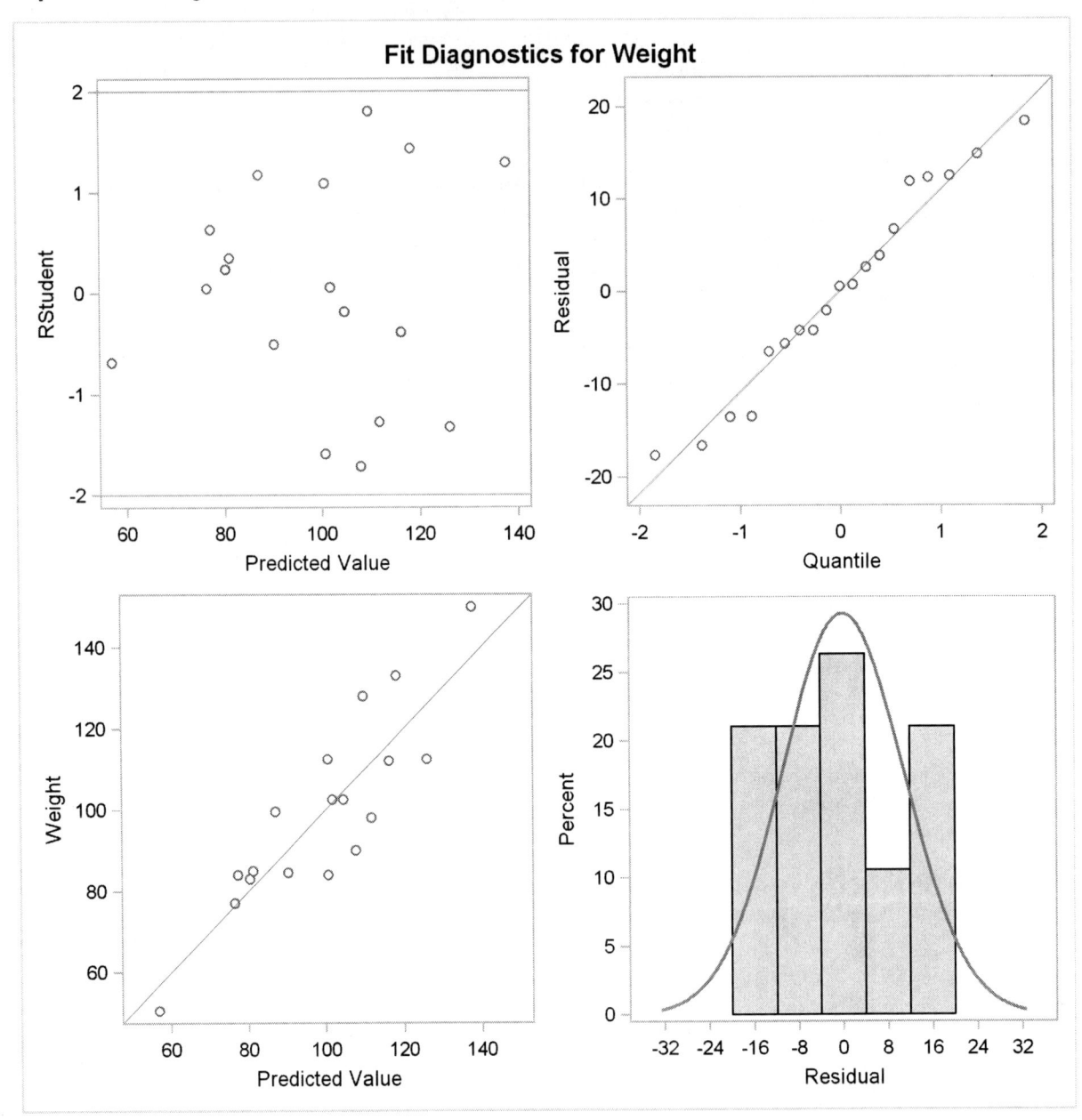

Fit Diagnostics for Weight

This new template is a straightforward modification of the original template. The COLUMNS=2 and ROWS=2 options in the LAYOUT LATTICE statement request a 2 × 2 lattice. The LAYOUT statement blocks for components 1, 3, 6, 8, and 9 are deleted. **NOTE:** You do not need to understand every aspect of a template to modify it if you can recognize the overall structure and a few key options.

You can restore the original template as follows:

```
proc template;
   delete Stat.REG.Graphics.DiagnosticsPanel;
run;
```

Example 21.8: Customizing Axes and Reference Lines

This example illustrates several ways that you can change the plot axes in a scatter plot. The example uses PROC CORRESP to perform a correspondence analysis. It is taken from the section "Getting Started: CORRESP Procedure" on page 1846 of Chapter 30, "The CORRESP Procedure." It uses the following data:

```
title "Number of Ph.D.'s Awarded from 1973 to 1978";

data PhD;
   input Science $ 1-19 y1973-y1978;
   label y1973 = '1973'
         y1974 = '1974'
         y1975 = '1975'
         y1976 = '1976'
         y1977 = '1977'
         y1978 = '1978';
   datalines;
Life Sciences       4489 4303 4402 4350 4266 4361
Physical Sciences   4101 3800 3749 3572 3410 3234
Social Sciences     3354 3286 3344 3278 3137 3008
Behavioral Sciences 2444 2587 2749 2878 2960 3049
Engineering         3338 3144 2959 2791 2641 2432
Mathematics         1222 1196 1149 1003  959  959
;
```

The following steps perform the correspondence analysis and create Output 21.8.1:

```
ods graphics on;
ods trace on;

proc corresp data=PhD short;
   ods select configplot;
   var y1973-y1978;
   id Science;
run;
```

Output 21.8.1 Default Scatter Plot

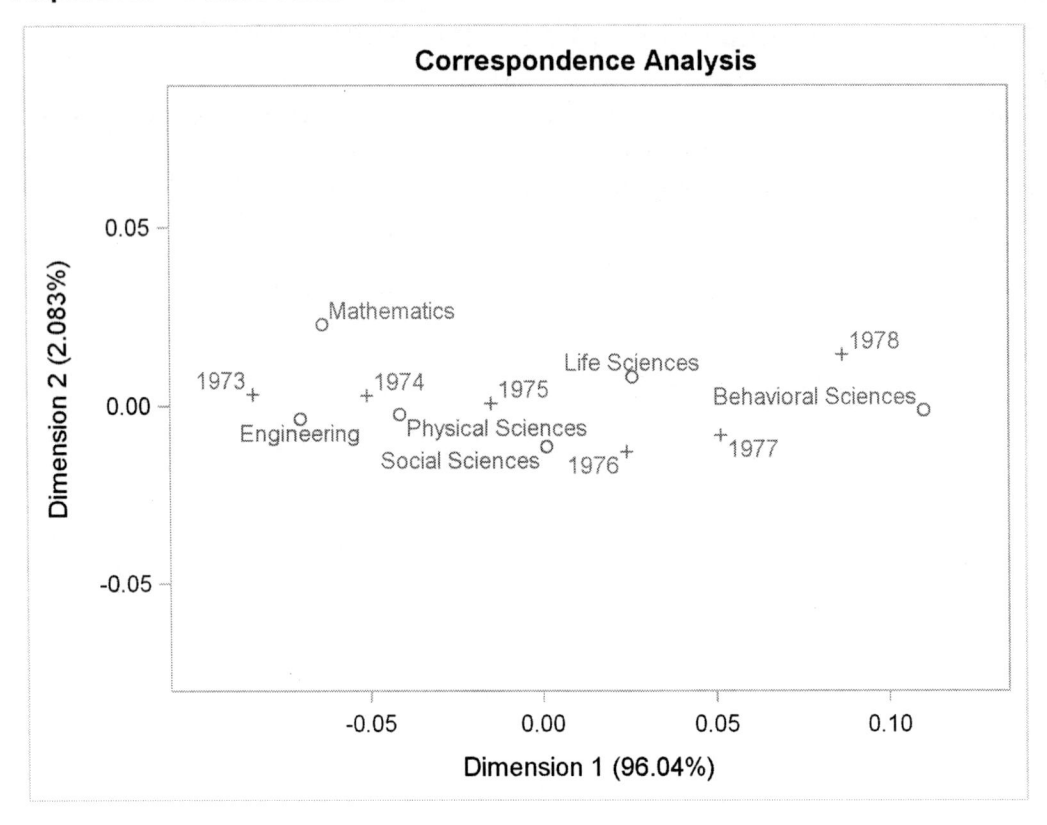

Correspondence Analysis

The trace output for this step (not shown) shows that the template for this plot is `Stat.Corresp.Graphics.Configuration`. The following step displays this template:

```
proc template;
   source Stat.Corresp.Graphics.Configuration;
run;
```

The results are as follows:

```
define statgraph Stat.Corresp.Graphics.Configuration;
   dynamic xVar yVar head legend;
   begingraph;
      entrytitle HEAD;
      layout overlayequated / equatetype=fit xaxisopts=(offsetmin=0.1
         offsetmax=0.1) yaxisopts=(offsetmin=0.1 offsetmax=0.1);
         scatterplot y=YVAR x=XVAR / group=GROUP index=INDEX
            datalabel=LABEL datalabelattrs=GRAPHVALUETEXT
            name="Type" tip=(y x datalabel group)
            tiplabel=(group="Point");
         if (LEGEND)
            discretelegend "Type";
         endif;
      endlayout;
   endgraph;
end;
```

You can add reference lines to the scatter plot at specified X and Y values by using the REFERENCELINE statement, as in the following example:

```
proc template;
   define statgraph Stat.Corresp.Graphics.Configuration;
      dynamic xVar yVar head legend;
      begingraph;
         entrytitle HEAD;
         layout overlayequated / equatetype=fit xaxisopts=(offsetmin=0.1
            offsetmax=0.1) yaxisopts=(offsetmin=0.1 offsetmax=0.1);

            referenceline x=0;
            referenceline y=0;

            scatterplot y=YVAR x=XVAR / group=GROUP index=INDEX
               datalabel=LABEL datalabelattrs=GRAPHVALUETEXT
               name="Type" tip=(y x datalabel group)
               tiplabel=(group="Point");
            if (LEGEND)
               discretelegend "Type";
            endif;
         endlayout;
      endgraph;
   end;
run;

proc corresp data=PhD short;
   ods select configplot;
   var y1973-y1978;
   id Science;
run;
```

When you modify templates, it is important to note that the order of the statements within the LAYOUT OVERLAYEQUATED (or more typically, the LAYOUT OVERLAY) is significant. Here, the reference lines are added before the scatter plot so that the reference lines are drawn before the scatter plot. Consequently, labels and markers that coincide with the reference lines are drawn over the reference lines. The results, with reference lines, are displayed in Output 21.8.2.

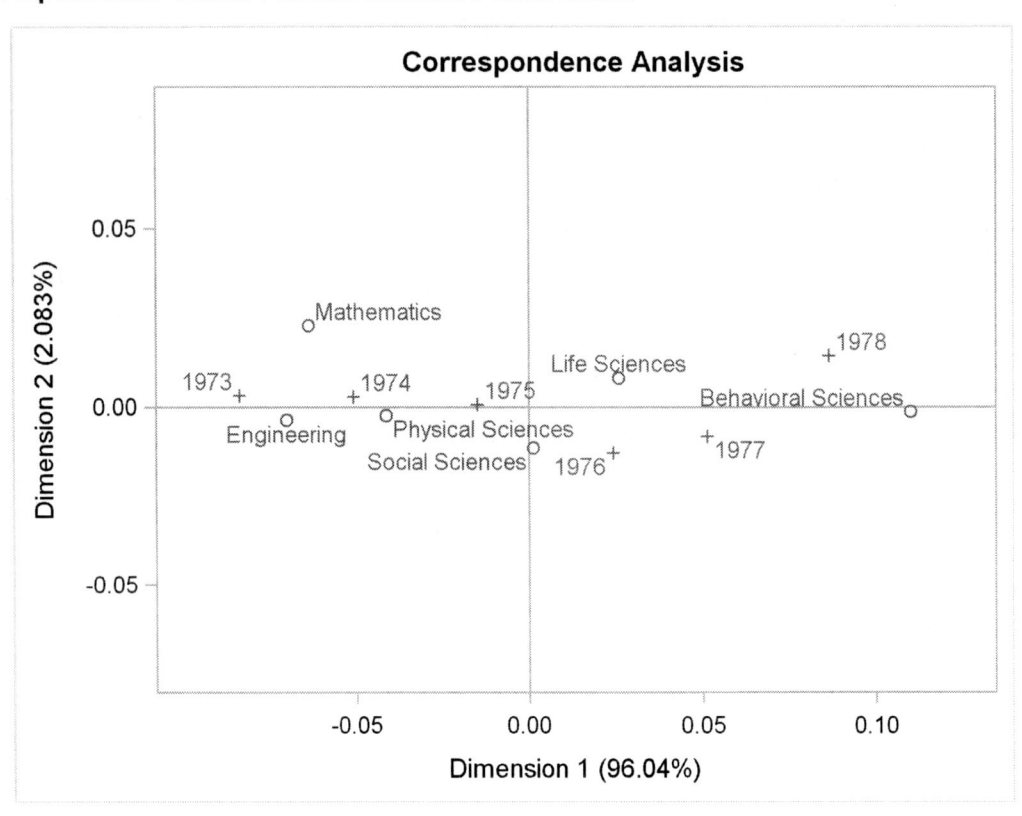

Output 21.8.2 Scatter Plot with Reference Lines Added

Correspondence Analysis scatter plot showing Dimension 1 (96.04%) vs Dimension 2 (2.083%), with points for Mathematics, Engineering, Physical Sciences, Social Sciences, Life Sciences, Behavioral Sciences and years 1973–1978.

You can restore the default graph template as follows:

```
proc template;
   delete Stat.Corresp.Graphics.Configuration;
run;
```

The next steps show how you can change the style so that a frame is not shown:

```
proc template;
   define style noframe;
      parent=styles.statistical;
      style graphwalls from graphwalls / frameborder=off;
   end;
run;

ods listing style=noframe;

proc corresp data=PhD short;
   ods select configplot;
   var y1973-y1978;
   id Science;
run;
```

The results, shown in Output 21.8.3, display an X-axis and a Y-axis without a frame. Unlike the previous change, which affects only the `ConfigPlot` display, this change affects all plots created with the NOFRAME style.

Output 21.8.3 Scatter Plot with No Axis Frame

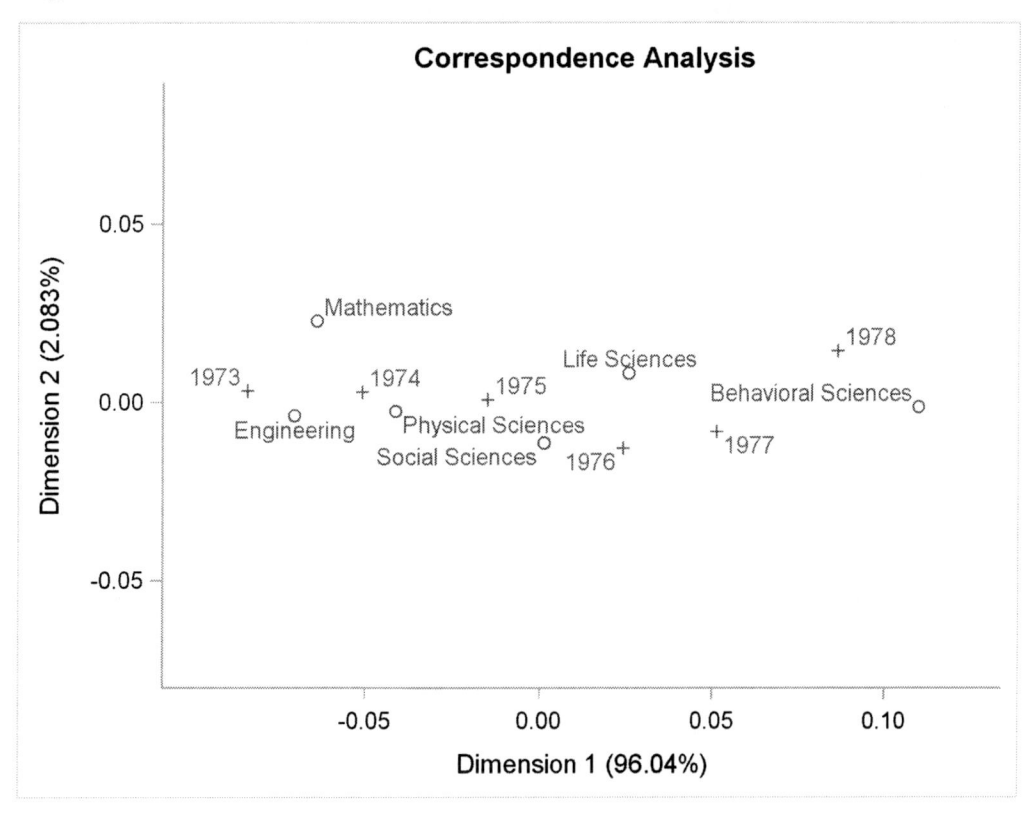

Correspondence Analysis

[Scatter plot showing Dimension 2 (2.083%) vs Dimension 1 (96.04%) with points labeled: Mathematics, 1973, 1974, 1975, Life Sciences, 1978, Engineering, Physical Sciences, Behavioral Sciences, Social Sciences, 1976, 1977]

Alternatively, you can also add reference lines and delete the entire axis frame using the WALLDISPLAY=NONE and the DISPLAY= option in the graph template, as in the following example:

```
proc template;
   define statgraph Stat.Corresp.Graphics.Configuration;
      dynamic xVar yVar head legend;
      begingraph;
         entrytitle HEAD;

         layout overlayequated / equatetype=fit walldisplay=none
            xaxisopts=(display=(tickvalues) offsetmin=0.1 offsetmax=0.1)
            yaxisopts=(display=(tickvalues) offsetmin=0.1 offsetmax=0.1);

            referenceline x=0;
            referenceline y=0;

            scatterplot y=YVAR x=XVAR / group=GROUP index=INDEX
               datalabel=LABEL datalabelattrs=GRAPHVALUETEXT
               name="Type" tip=(y x datalabel group)
               tiplabel=(group="Point");
            if (LEGEND)
               discretelegend "Type";
            endif;
         endlayout;
      endgraph;
   end;
run;
```

```
ods listing style=statistical;

proc corresp data=PhD short;
   ods select configplot;
   var y1973-y1978;
   id Science;
run;
```

The results are shown in Output 21.8.4.

Output 21.8.4 Scatter Plot with Internal Axes

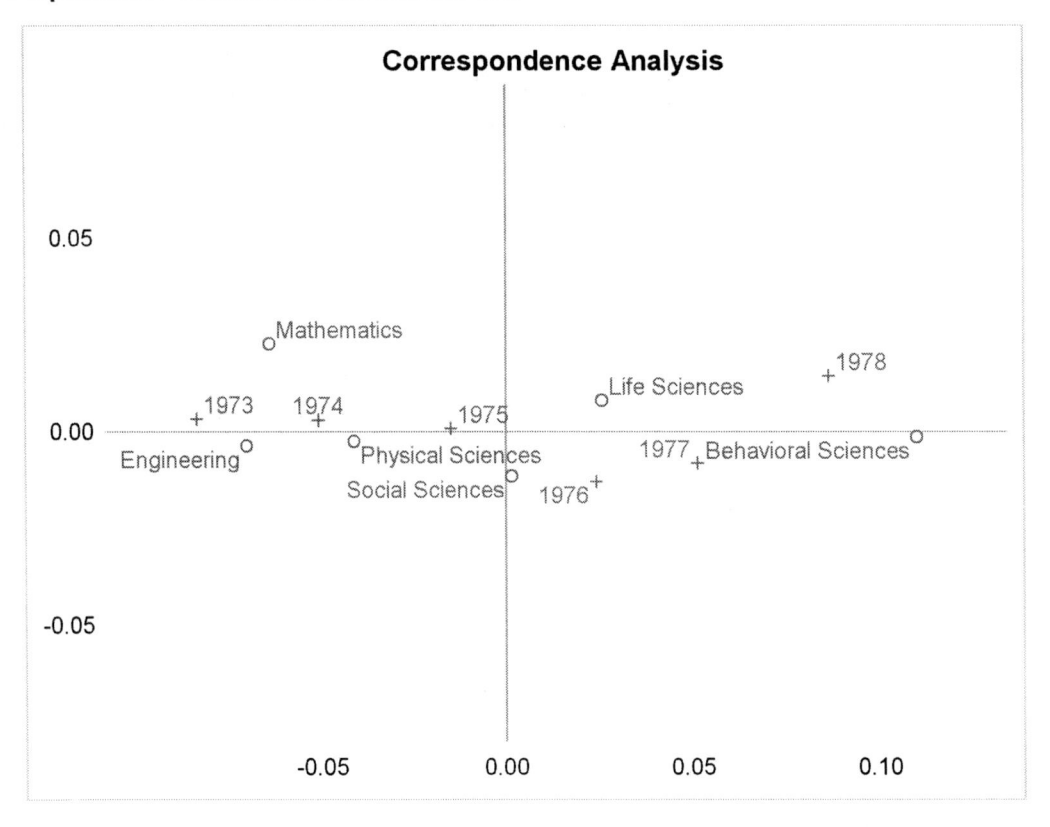

Instead of DISPLAY=(TICKVALUES), you can use DISPLAY=NONE (not shown) to remove the tick values from the display as well. You can change the tick values, as in the following example:

```
proc template;
   define statgraph Stat.Corresp.Graphics.Configuration;
      dynamic xVar yVar head legend;
      begingraph;
         entrytitle HEAD;

         layout overlayequated / equatetype=fit
            commonaxisopts=(tickvaluelist=(0))
            xaxisopts=(offsetmin=0.1 offsetmax=0.1)
            yaxisopts=(offsetmin=0.1 offsetmax=0.1);

            referenceline x=0;
            referenceline y=0;
```

```
            scatterplot y=YVAR x=XVAR / group=GROUP index=INDEX
               datalabel=LABEL datalabelattrs=GRAPHVALUETEXT
               name="Type" tip=(y x datalabel group)
               tiplabel=(group="Point");
            if (LEGEND)
               discretelegend "Type";
            endif;
         endlayout;
      endgraph;
   end;
run;

proc corresp data=PhD short;
   ods select configplot;
   var y1973-y1978;
   id Science;
run;
```

Since the axes in this plot are equated, the ticks are specified using the option `commonaxisopts=(tickvaluelist=(tick-value-list))`. This example only shows ticks at zero, but you can specify lists of values instead. The results are shown in Output 21.8.5.

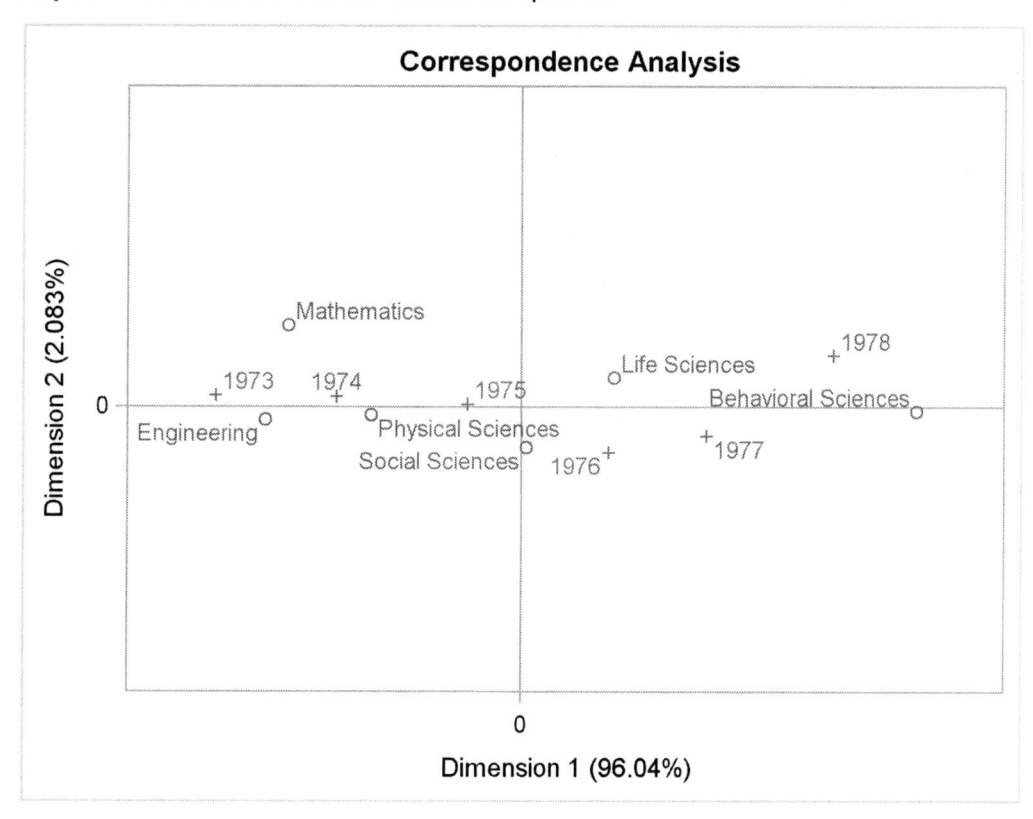

Output 21.8.5 Scatter Plot with Tick Marks Specified

If the axes are not equated, then the tick value list is specified with the LINEAROPTS= option, as in the following statement:

```
layout overlay / xaxisopts=(linearopts=(viewmin=-0.1 viewmax=0.1
                                       tickvaluelist=(-0.1 0 0.1))
                            offsetmin=0.1 offsetmax=0.1)
                 yaxisopts=(linearopts=(viewmin=-0.1 viewmax=0.1
                                       tickvaluelist=(-0.1 0 0.1))
                            offsetmin=0.1 offsetmax=0.1);
```

The preceding statement uses the VIEWMIN= and VIEWMAX= options to specify the beginning and end of the data range that is shown. Specifying a tick value list does not extend or restrict the range of data shown in the plot. When axes share common options, it might be more convenient to use a macro to specify the options. The following two statements are equivalent to the preceding statement:

```
%let opts = linearopts=(viewmin=-0.1 viewmax=0.1
             tickvaluelist=(-0.1 0 0.1)) offsetmin=0.1 offsetmax=0.1;
layout overlay / xaxisopts=(&opts) yaxisopts=(&opts);
```

You can restore the default graph template as follows:

```
proc template;
    delete Stat.Corresp.Graphics.Configuration;
run;
```

Example 21.9: Customizing the Style for Box Plots

This example demonstrates how to modify the style for box plots. This example is taken from Example 21.1. The following step creates the data set:

```
data pr;
    input Person Gender $ y1 y2 y3 y4 @@;
    y=y1; Age=8;  output;
    y=y2; Age=10; output;
    y=y3; Age=12; output;
    y=y4; Age=14; output;
    drop y1-y4;
    datalines;
 1  F   21.0  20.0  21.5  23.0    2  F   21.0  21.5  24.0  25.5

    ... more lines ...

;
```

The next step displays the STATISTICAL and DEFAULT styles:

```
proc template;
   source Styles.Statistical;
   source Styles.Default;
run;
```

If you search for 'box', you will find the style element that controls some aspects of the box plot:

```
class GraphBox /
   capstyle = "serif"
   connect = "mean"
   displayopts = "fill caps median mean outliers";
```

You can learn more about the `GraphBox` style element and its attributes in the section on the BOXPLOT statement in the *SAS/GRAPH: Graph Template Language Reference* and in the section on "ODS Style Elements" in the *SAS Output Delivery System: User's Guide*.

The following statements create two new styles by modifying attributes of the `GraphBox` style element. The first style is a sparse style; the box is outlined (not filled) and the median is shown but not the mean. In contrast, the second style produces a filled box, with caps on the whiskers that shows the mean, median, and outliers. In addition, the box is notched. The following statements create the two styles:

```
proc template;
   define style boxstylesparse;
      parent=styles.statistical;
      style GraphBox / capstyle = "line" displayopts = "median";
   end;
   define style boxstylerich;
      parent=styles.statistical;
      style GraphBox / capstyle = "bracket"
              displayopts = "fill caps median mean outliers notches";
   end;
run;
```

The following steps run PROC MIXED and create box plots that use the two styles:

```
ods graphics on;
ods listing style=boxstylesparse;

proc mixed data=pr method=ml plots=boxplot;
   ods select 'Conditional Residuals by Gender';
   class Person Gender;
   model y = Gender Age Gender*Age;
   random intercept Age / type=un subject=Person;
run;
```

```
ods listing style=boxstylerich;

proc mixed data=pr method=ml plots=boxplot;
   ods select 'Conditional Residuals by Gender';
   class Person Gender;
   model y = Gender Age Gender*Age;
   random intercept Age / type=un subject=Person;
run;
```

The results with the sparse style are displayed in Output 21.9.1, and the results with the richer style are displayed in Output 21.9.2. See Output 21.1.1 in Example 21.1 to see the results using the STATISTICAL style.

Output 21.9.1 Box Plot with the Sparse Style

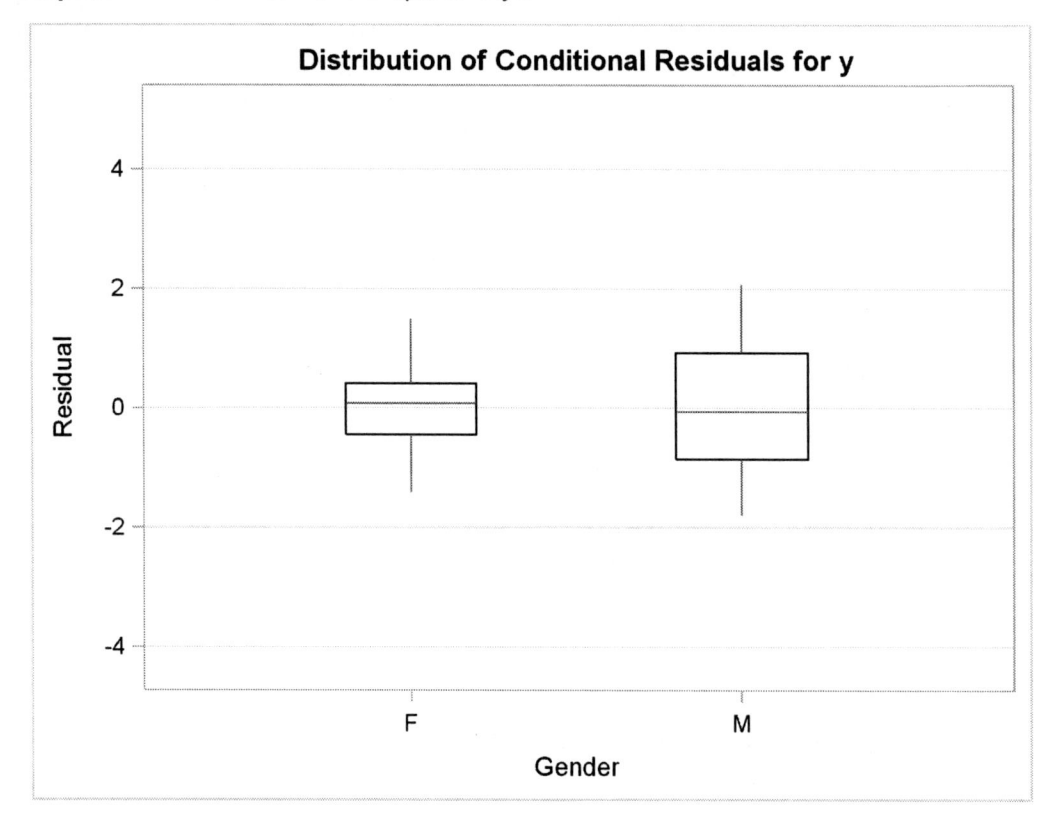

Output 21.9.2 Box Plot with the Richer Style

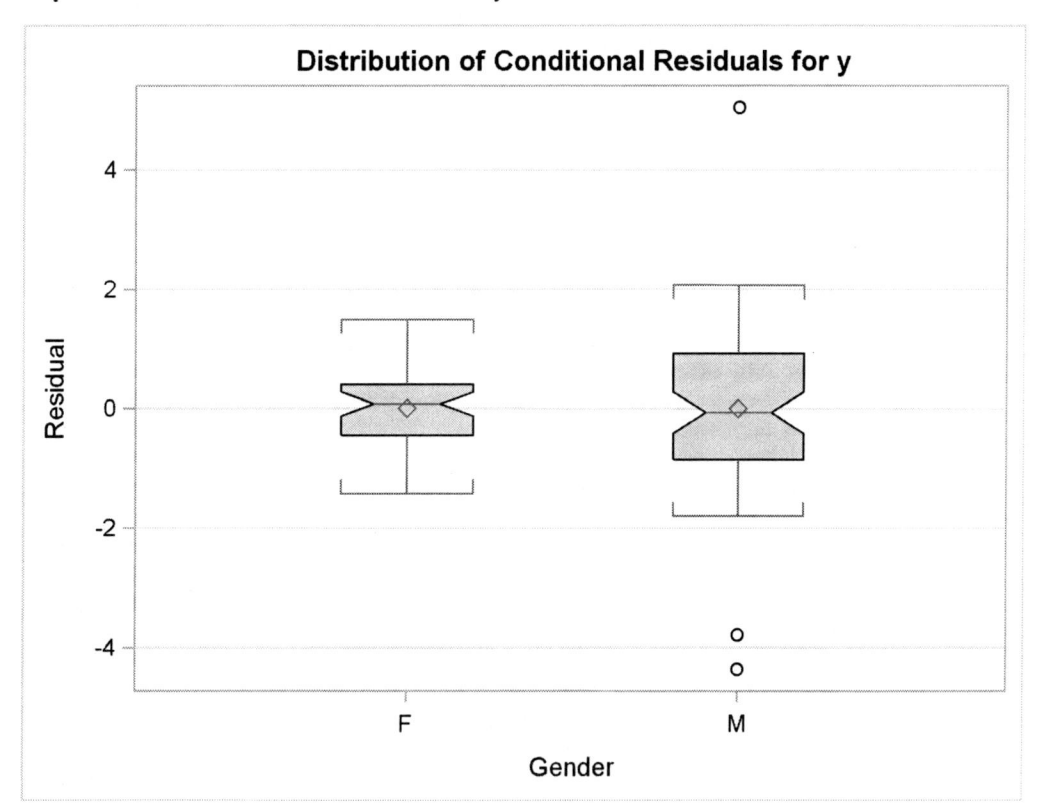

Example 21.10: Adding Text to Every Graph

This example shows how to add text to one or more graphs. For example, you can create a macro variable, with project and date information, as follows:

```
%let date = Project 17.104, &sysdate;
```

In order to add this information to a set of graphs, you need to first know the names of their templates. You can list the names of every graph template for SAS/STAT procedures or for a particular procedure as follows:

```
proc template;
   list stat     / where=(type='Statgraph');
   list stat.reg / where=(type='Statgraph');
run;
```

The results for PROC REG are shown in Output 21.10.1.

Output 21.10.1 PROC REG Templates

```
        Listing of: SASHELP.TMPLMST
        Path Filter is: Stat.Reg
        Sort by: PATH/ASCENDING

        Obs     Path                                              Type
        ----------------------------------------------------------------------
         1      Stat.Reg.Graphics.CooksD                          Statgraph
         2      Stat.Reg.Graphics.DFBETASPanel                    Statgraph
         3      Stat.Reg.Graphics.DFBETASPlot                     Statgraph
         4      Stat.Reg.Graphics.DFFITSPlot                      Statgraph
         5      Stat.Reg.Graphics.DiagnosticsPanel                Statgraph
         6      Stat.Reg.Graphics.Fit                             Statgraph
         7      Stat.Reg.Graphics.ObservedByPredicted             Statgraph
         8      Stat.Reg.Graphics.PartialPanel                    Statgraph
         9      Stat.Reg.Graphics.PartialPlot                     Statgraph
        10      Stat.Reg.Graphics.PredictionPanel                 Statgraph
        11      Stat.Reg.Graphics.QQPlot                          Statgraph
        12      Stat.Reg.Graphics.RFPlot                          Statgraph
        13      Stat.Reg.Graphics.RStudentByPredicted             Statgraph
        14      Stat.Reg.Graphics.ResidualBoxPlot                 Statgraph
        15      Stat.Reg.Graphics.ResidualByPredicted             Statgraph
        16      Stat.Reg.Graphics.ResidualHistogram               Statgraph
        17      Stat.Reg.Graphics.ResidualPanel                   Statgraph
        18      Stat.Reg.Graphics.ResidualPlot                    Statgraph
        19      Stat.Reg.Graphics.RidgePanel                      Statgraph
        20      Stat.Reg.Graphics.RidgePlot                       Statgraph
        21      Stat.Reg.Graphics.SelectionCriterionPanel         Statgraph
        22      Stat.Reg.Graphics.SelectionCriterionPlot          Statgraph
        23      Stat.Reg.Graphics.StepSelectionCriterionPanel     Statgraph
        24      Stat.Reg.Graphics.StepSelectionCriterionPlot      Statgraph
        25      Stat.Reg.Graphics.VIFPlot                         Statgraph
        26      Stat.Reg.Graphics.rstudentByLeverage              Statgraph
```

You can show the source for the graph templates for SAS/STAT procedures or for a particular procedure as follows:

```
options ls=96;
proc template;
   source stat     / where=(type='Statgraph');
   source stat.reg / where=(type='Statgraph');
options ls=80;
```

The results of this step are not shown. However, Example 21.7 shows a portion of the template for the PROC REG diagnostics panel. Here, the OPTIONS statement is used to set a line size of 96, which sometimes works better than the smaller default line sizes when showing the source for large and complicated templates.

An abridged version of the first few lines of the diagnostics panel template is displayed next:

```
define statgraph Stat.Reg.Graphics.DiagnosticsPanel;
   notes "Diagnostics Panel";
   dynamic . . .;
   BeginGraph / designheight=defaultDesignWidth;
      entrytitle halign=left textattrs=GRAPHVALUETEXT _MODELLABEL
         halign=center textattrs=GRAPHTITLETEXT "Fit Diagnostics"
         " for " _DEPNAME;
      . . .
```

Adding a Date and Project Stamp to a Few Graphs

You can add the project and date to the bottom of all graphs produced with PROC REG by putting a PROC TEMPLATE statement in front of the template source code, and adding an MVAR and ENTRYFOOTNOTE statement after every BEGINGRAPH statement, as in the following example:

```
proc template;
   define statgraph Stat.Reg.Graphics.DiagnosticsPanel;
      notes "Diagnostics Panel";
      dynamic . . .;
      BeginGraph / designheight=defaultDesignWidth;

         mvar date;
         entryfootnote halign=left textattrs=GraphValueText date;

         entrytitle halign=left textattrs=GRAPHVALUETEXT _MODELLABEL
            halign=center textattrs=GRAPHTITLETEXT "Fit Diagnostics"
            " for " _DEPNAME;
         . . .
```

The MVAR statement enables you to dynamically customize the template and graph at procedure run time, just as the DYNAMIC statement enables the procedure to dynamically customize the template and graph. With the MVAR statement, you can modify the template once and reuse that modification as the macro changes over time: Alternatively, you can modify the templates as follows:

```
entryfootnote halign=left textattrs=GRAPHVALUETEXT "&date";
```

However, you would then have to resubmit your templates every time the macro variable changed. The substitution for the macro variable date occurs at different times in the two preceding cases. In the former case, ODS looks for the value of the macro variable date at the time the template is used, and then the current date variable is used to set the text in the ENTRYFOOTNOTE statement, every time the template is used. In the latter case, SAS substitutes the value of the macro variable once, at the time that the PROC TEMPLATE step is executed.

The following steps use the Class data set and produce Output 21.10.2:

Output 21.10.2 PROC REG Plots with Project and Date Stamp

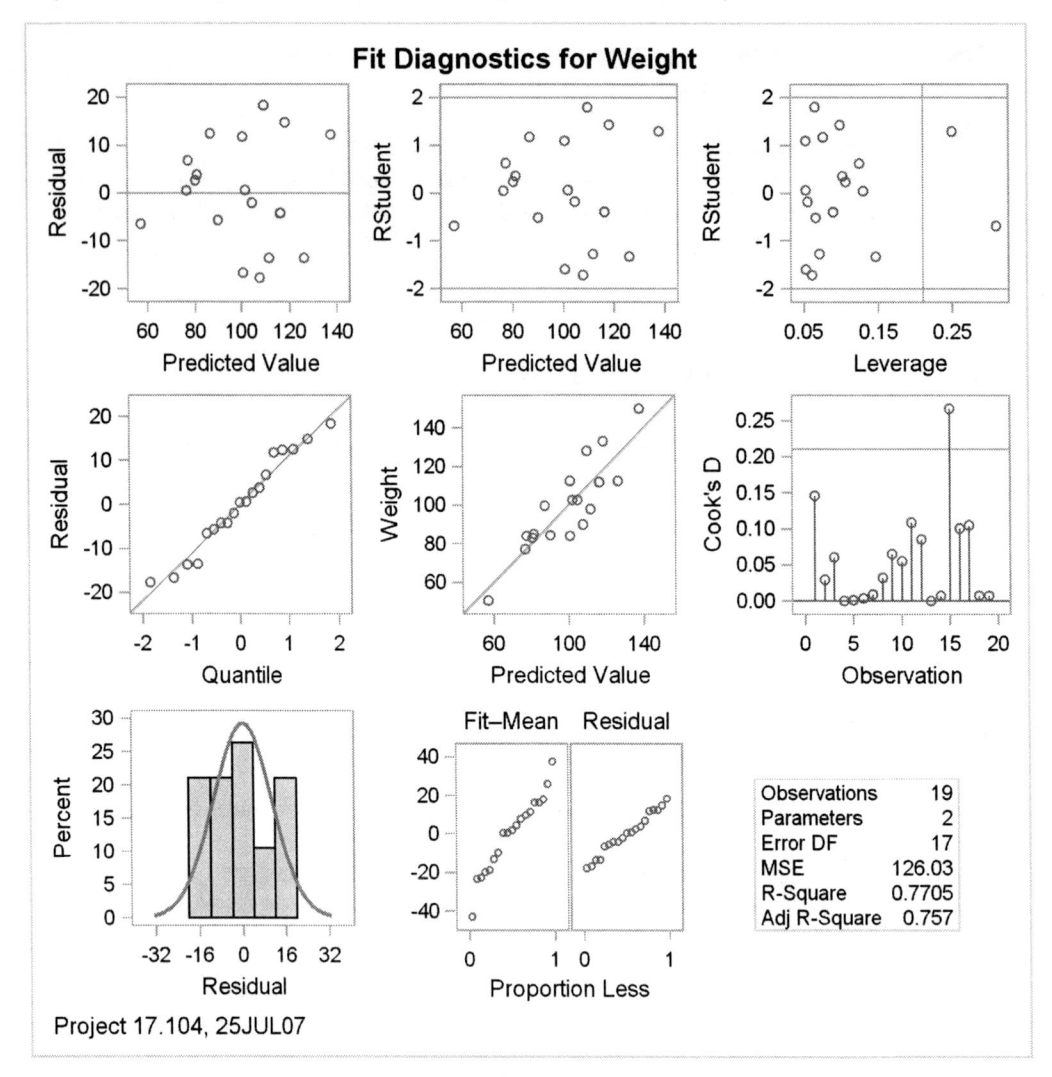

Fit Diagnostics for Weight

Project 17.104, 25JUL07

Observations	19
Parameters	2
Error DF	17
MSE	126.03
R-Square	0.7705
Adj R-Square	0.757

Output 21.10.2 *continued*

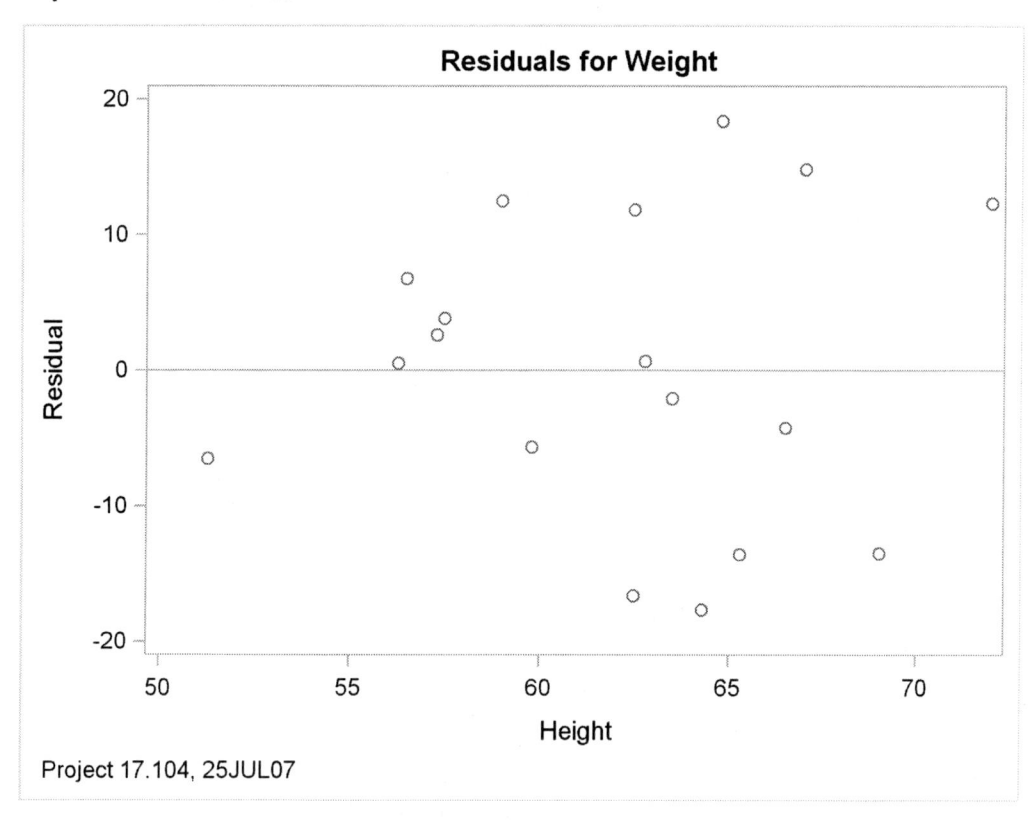

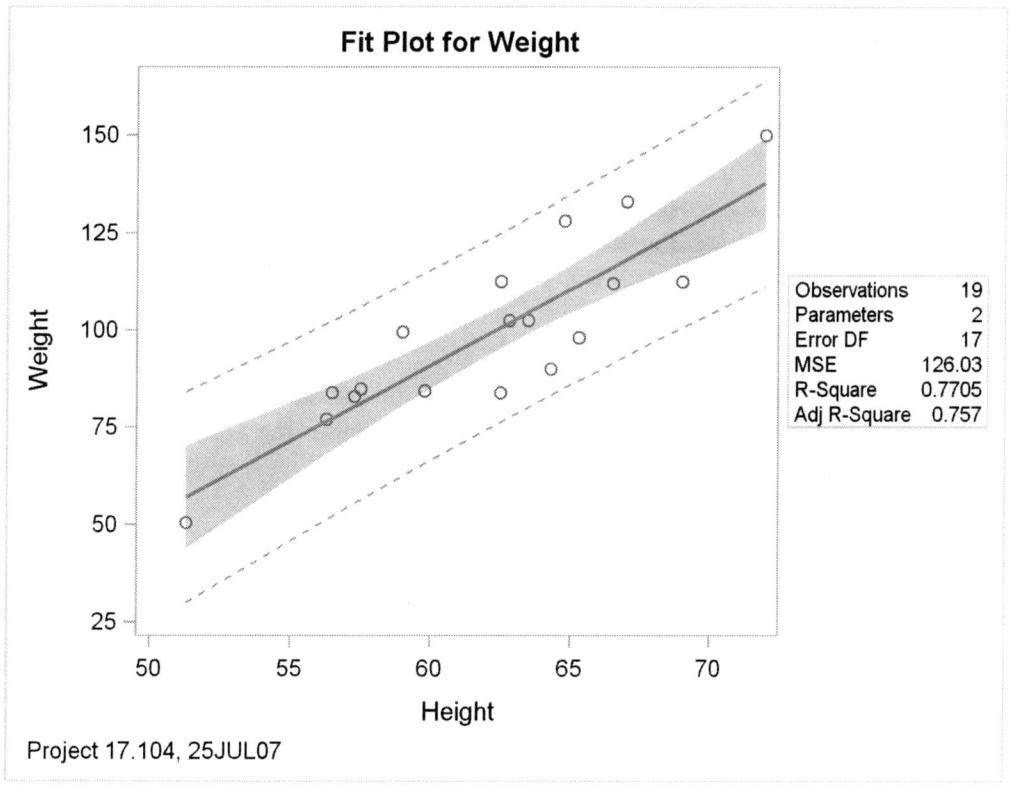

Residuals for Weight

Project 17.104, 25JUL07

Fit Plot for Weight

Observations	19
Parameters	2
Error DF	17
MSE	126.03
R-Square	0.7705
Adj R-Square	0.757

Project 17.104, 25JUL07

You can restore all of the default templates for PROC REG by running the following step:

```
proc template;
   delete stat.reg;
run;
```

Alternatively, you can specify `delete stat` to restore all SAS/STAT templates to their default definitions.

You can add text to the top or the bottom of a graph by using the ENTRYTITLE or the ENTRYFOOTNOTE statement, respectively. With both statements, you can put the text in the HALIGN=RIGHT, HALIGN=LEFT, or HALIGN=CENTER positions. You can add text to titles even if they already have a centered title. For example, the ENTRYTITLE statement in the diagnostic panel has text on the left (which is conditionally displayed) and a centered title:

```
entrytitle halign=left textattrs=GraphValueText _MODELLABEL
   halign=center textattrs=GraphTitleText "Fit Diagnostics"
   " for " _DEPNAME;
```

The current title can be followed by HALIGN=RIGHT and more text.

Adding Data Set Information to a Graph

You might, for example, want to add text to a set of graphs that indicates the most recently created data set. The following example shows you how you can do this with the syslast macro variable:

```
%let data = &syslast;

. . .

mvar data;
entrytitle halign=left textattrs=GraphValueText "Data: " data
   halign=center textattrs=GraphTitleText "Fit Diagnostics"
   " for " _DEPNAME;

. . .
```

Of course, this only makes sense when you are analyzing the last data set created. Alternatively, you can incorporate the name of the data set in the title, as in the following example:

```
%let data = &syslast;

. . .

mvar data;
entrytitle halign=center textattrs=GraphTitleText
         "Fit Diagnostics for Data Set " data;

. . .
```

Adding a Date and Project Stamp to All Graphs

Sometimes, you can automate the process of template modification. For example, you can automatically add an MVAR and ENTRYFOOTNOTE statement to every graph template, as in the following example:

```
options ls=256;

proc template;
   source / where=(type='Statgraph') file="tpls.sas";
run;

options ls=80;

data _null_;
   infile 'tpls.sas' lrecl=256 pad;
   input line $ 1-256;
   file 'newtpls.sas';
   put line;
   line = left(lowcase(line));
   if line =: 'begingraph' then
      put 'mvar __date;' /
          'entryfootnote halign=left textattrs=GraphValueText __date;';

   file log;
   if index(line, '__date') then
      put 'ERROR: Name __date already used.' / line;
   if index(line, 'entryfootnote') then put line;
run;

proc template;
   %include 'newtpls.sas' / nosource;
run;
```

These statements write all ODS graph templates to a file, read that file, and write out a new file with an MVAR and ENTRYFOOTNOTE statement added after every BEGINGRAPH statement. Then these new templates are compiled with PROC TEMPLATE. These steps assume that no BEGINGRAPH statement is longer than 256 characters. Most graphs do not have footnotes. Those that do will now have multiple footnotes. You might want to manually combine them or write a more complicated program to handle them. These steps also assume that the name __date is not used anywhere. However, the program does check this and also lists all ENTRYFOOTNOTE statements. Be careful to check the SAS log to ensure that all templates compile without error. Also, before using templates that are automatically modified, make sure your modifications are reasonable.

You can delete SASUSER.Templat and hence all modified templates (assuming the default template search path) as follows:

```
ods path sashelp.tmplmst(read);
proc datasets library=sasuser nolist;
   delete templat(memtype=itemstor);
run;
ods path sasuser.templat(update) sashelp.tmplmst(read);
```

Example 21.11: PROC TEMPLATE Statement Order and Primary Plots

This example uses artificial data to illustrate two basic principles of template writing: that statement order matters and that one of the plotting statements is the primary statement. The data are a sample from a bivariate normal distribution. A custom graph template and PROC SGRENDER are used to plot the data along with vectors and ellipses. The plot consists of four components: a scatterplot of the data; vectors whose end points come from other variables in the data set; ellipses whose parameters are specified in the template; and reference lines whose locations are specified in the template. Initially, thick lines are used to show what happens at the places where the lines and points intersect.

The following steps create the input SAS data set:

```
data x;
   input x y;
   label x = 'Normal(0, 4)' y = 'Normal(0, 1)';
   datalines;
-4  0
 4  0
 0 -2
 0  2
;

data y(drop=i);
   do i = 1 to 2500;
      r1 = normal( 104 );
      r2 = normal( 104 ) * 2;
      output;
   end;
run;

data all;
   merge x y;
run;
```

The data set All contains four variables. The variables r1 and r2 contain the random data. These variables contain 2500 nonmissing observations. The data set also contains the variables x and y, which contain the end points for the vectors. These variables contain four nonmissing observations and 2496 observations that are all missing. A data set like this is not unusual when creating overlaid plots. Different overlays often require input data with very different sizes. First, the data are plotted by using a template that is deliberately constructed to demonstrate a number of problems that can occur with statement order.

Example 21.11: PROC TEMPLATE Statement Order and Primary Plots ✦ 773

The following steps create Output 21.11.1:

```
proc template;
   define statgraph Plot;
      begingraph;
         entrytitle 'Statement Order and the PRIMARY= Option';
         layout overlayequated / equatetype=fit;
            ellipseparm semimajor=eval(sqrt(4)) semiminor=1
                        slope=0 xorigin=0 yorigin=0 /
                        outlineattrs=GraphData2(pattern=solid thickness=5);
            ellipseparm semimajor=eval(2 * sqrt(4)) semiminor=2
                        slope=0 xorigin=0 yorigin=0 /
                        outlineattrs=GraphData5(pattern=solid thickness=5);
            vectorplot y=y x=x xorigin=0 yorigin=0 /
                        arrowheads=false lineattrs=GraphFit(thickness=5);
            scatterplot y=r1 x=r2 /
                        markerattrs=(symbol=circlefilled size=3);
            referenceline x=0 / lineattrs=(thickness=3);
            referenceline y=0 / lineattrs=(thickness=3);
         endlayout;
      endgraph;
   end;
run;

ods listing style=listing;

proc sgrender data=all template=plot;
run;
```

Output 21.11.1 Statements Specified in a Nonoptimal Order

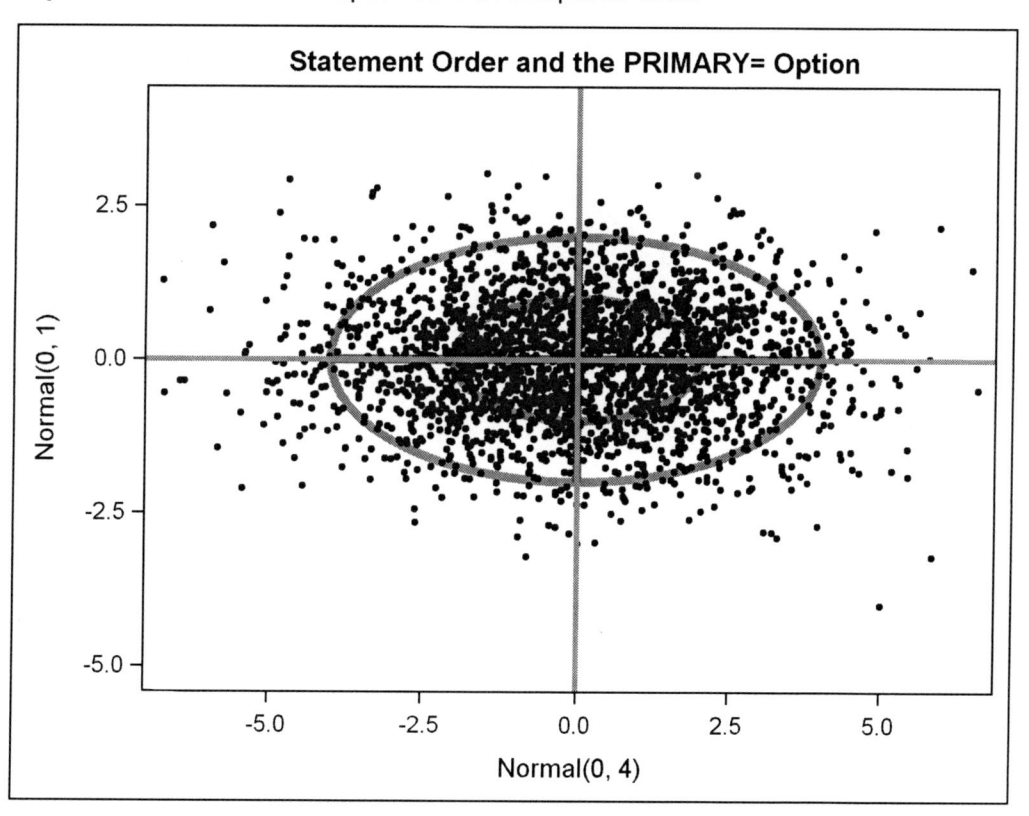

There are a number of problems with the plot in Output 21.11.1. The reference lines obliterate the vectors, and the data are on top of everything but the reference lines. It might be more reasonable to plot the reference lines first, the data next, the vectors next, and the ellipses last. The following steps do this and produce Output 21.11.2:

```
proc template;
   define statgraph Plot;
      begingraph;
         entrytitle 'Statement Order and the PRIMARY= Option';
         layout overlayequated / equatetype=fit;
            referenceline x=0 / lineattrs=(thickness=3);
            referenceline y=0 / lineattrs=(thickness=3);
            scatterplot y=r1 x=r2 /
                        markerattrs=(symbol=circlefilled size=3);
            vectorplot  y=y x=x xorigin=0 yorigin=0 /
                        arrowheads=false lineattrs=GraphFit(thickness=5);
            ellipseparm semimajor=eval(sqrt(4)) semiminor=1
                        slope=0 xorigin=0 yorigin=0 /
                        outlineattrs=GraphData2(pattern=solid thickness=5);
            ellipseparm semimajor=eval(2 * sqrt(4)) semiminor=2
                        slope=0 xorigin=0 yorigin=0 /
                        outlineattrs=GraphData5(pattern=solid thickness=5);
         endlayout;
      endgraph;
   end;
run;

ods listing style=listing;

proc sgrender data=all template=plot;
run;
```

Output 21.11.2 Statement Order Fixed

Statement Order and the PRIMARY= Option

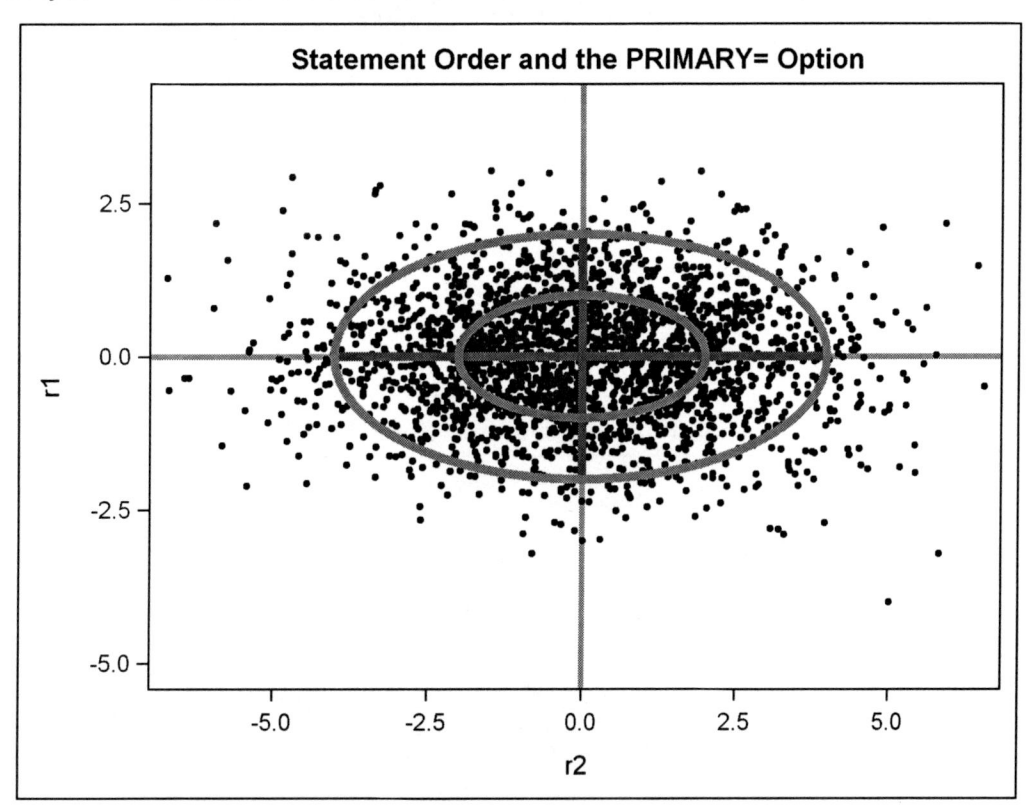

Output 21.11.2 looks better than Output 21.11.1, but the labels for the axes have changed. Output 21.11.1 has the labels of the variables x and y as axis labels, whereas Output 21.11.2 uses the names of the variables r1 and r2. This is because in the Output 21.11.1, the first plot is the vector plot of x and y (which have labels), and in Output 21.11.2, the first plot is the scatter plot of r1 and r2 (which do not have labels). By default, the first plot is the *primary plot*, and the primary plot is used to determine the axis type and labels. You can designate the vector plot as the primary plot with the PRIMARY=TRUE option. The following statements make the final plot, this time with default line thicknesses, and produce Output 21.11.3:

```
proc template;
   define statgraph Plot;
      begingraph;
         entrytitle 'Statement Order and the PRIMARY= Option';
         layout overlayequated / equatetype=fit;
            referenceline x=0;
            referenceline y=0;
            scatterplot y=r1 x=r2 / markerattrs=(symbol=circlefilled size=3);

            vectorplot  y=y x=x xorigin=0 yorigin=0 / primary=true
                        arrowheads=false lineattrs=GraphFit;

            ellipseparm semimajor=eval(sqrt(4)) semiminor=1
                        slope=0 xorigin=0 yorigin=0 /
                        outlineattrs=GraphData2(pattern=solid);
            ellipseparm semimajor=eval(2 * sqrt(4)) semiminor=2
```

```
                    slope=0 xorigin=0 yorigin=0 /
                    outlineattrs=GraphData5(pattern=solid);
            endlayout;
         endgraph;
      end;
run;

ods listing style=listing;

proc sgrender data=all template=plot;
run;
```

Output 21.11.3 Statement Order Fixed and Primary Plot Specified

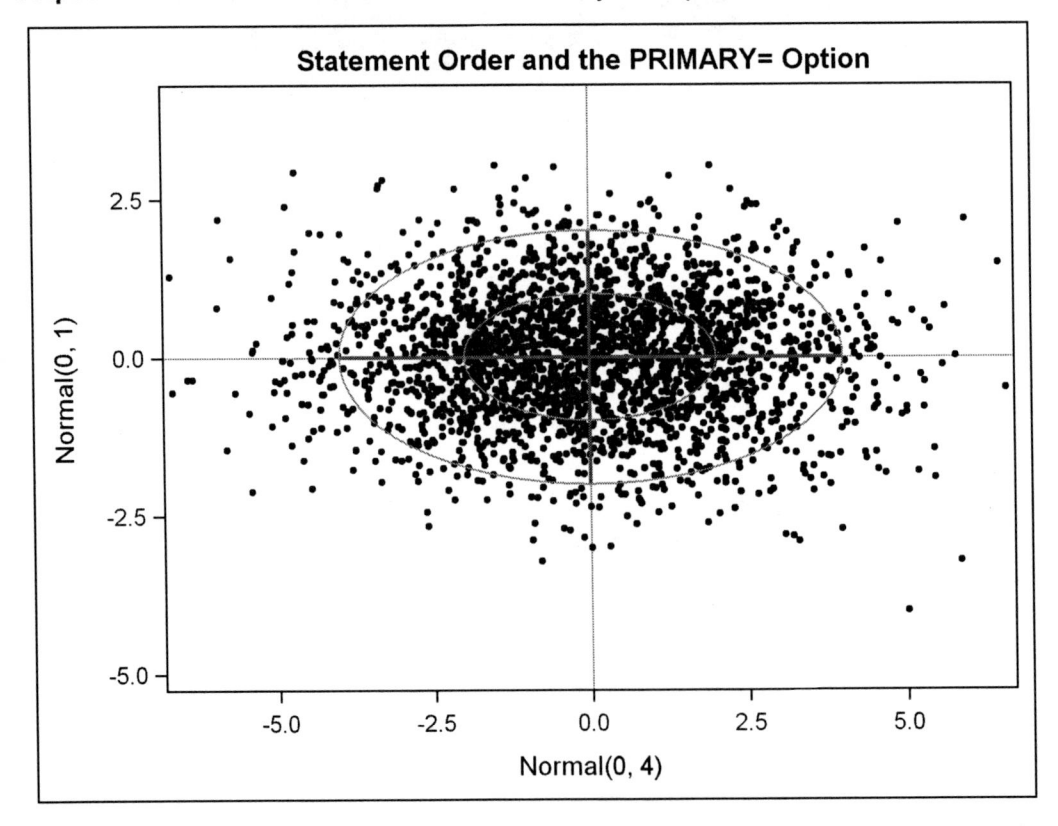

The axis labels in Output 21.11.3 and the overprinting of plot elements look better than in the previous plots. You can further adjust the line thicknesses if you want to emphasize or deemphasize components of this plot. The following list discusses the syntax of the GTL statements used in this example.

- The template has an ENTRYTITLE statement that specifies the title.

- The template has an equated overlay. This means that a centimeter on one axis represents the same data range as a centimeter on the other axis. This is done instead of the more common LAYOUT OVERLAY since with these data, the shape and geometry of the data have meaning even though the ranges of the two axis variables are different. The option EQUATETYPE=SQUARE is used to make a square plot, but since the X-axis variable has a larger range than the Y-axis variable, and since the default plot size is wider than high, EQUATETYPE=FIT is specified. The axes are equated but use the available space.

- A vertical reference line is drawn at X=0, and a horizontal reference line is drawn at Y=0.

- The scatter plot is based on the Y-axis variable r2 and the X-axis variable r1. The markers are filled circles with a size of three pixels. This is smaller than the default size and works well with a plot that displays many points.

- The vector plot is based on the Y-axis variable y and the X-axis variable x. The vectors are solid lines with no heads emanating from the origin (X=0 and Y=0). The color and other line attributes such as thickness come from the attributes of the `GraphFit` style element. This is the primary plot, so the default axis labels are the variable labels for the X= and Y= variables if they exist or the variable names if the variables do not have labels.

- The plot also displays two ellipses with X=0 and Y=0 at their center. Their widths are expressions, and their heights are constant. The expressions are not needed in this example; they are used to illustrate the syntax. The SEMIMAJOR= option specifies half the length of the major axis for the ellipse, and the SEMIMINOR= option specifies half the length of the minor axis for the ellipse. The SLOPE= option specifies the slope of the major axis for the ellipse. The colors of the ellipses and other line properties are based on the `GraphData2` and `GraphData5` style elements, but the line pattern attribute from the style is overridden.

References

Kuhfeld, W. F. (2009), "Modifying ODS Statistical Graphics Templates in SAS 9.2," `http://support.sas.com/rnd/app/papers/modtmplt.pdf`.

Subject Index

ANALYSIS style
    ODS Styles, 618
axis customization
    ODS Graphics, 755

box plot
    ODS Graphics, 719, 763

DEFAULT style
    ODS Styles, 617
destination, closing
    ODS Graphics, 656
destinations
    ODS Graphics, 651
diagnostics panel
    ODS Graphics, 751
DOCUMENT destination
    ODS Graphics, 724
document path
    ODS Graphics, 726
DOCUMENT procedure
    document path, 726
    ODS Graphics, 724
Documents window
    ODS Graphics, 725

fonts, modifying
    ODS Graphics, 680

graph label
    ODS Graphics, 647
graph modification
    ODS Graphics, 634
graph name
    ODS Graphics, 647, 725
graph resolution
    ODS Graphics, 633, 656
graph size
    ODS Graphics, 633, 656
graph template language
    ODS Graphics, 687
graph templates
    ODS Graphics, 687
graph templates, customizing
    ODS Graphics, 697
graph templates, definition
    ODS Graphics, 730
graph templates, displaying
    ODS Graphics, 693

graph templates, editing
    ODS Graphics, 695
graph templates, locating
    ODS Graphics, 692
graph templates, reverting to default
    ODS Graphics, 698
graph templates, saving
    ODS Graphics, 697, 731
graph titles, modifying
    ODS Graphics, 731
graphics image file
    file type, 651
    ODS Graphics, 651–653
    PostScript, 655, 722
graphics image file, saving
    ODS Graphics, 654
graphics image file, type
    ODS Graphics, 651
grid lines
    ODS Graphics, 737

histogram
    ODS Graphics, 722
HTML destination
    ODS Graphics, 651, 654

index counter
    ODS Graphics, 652

JOURNAL style
    ODS Styles, 722
JOURNAL2 style
    ODS Styles, 620

LaTeX destination
    ODS Graphics, 651, 655, 722
LISTING destination
    ODS Graphics, 654
LISTING style
    ODS Styles, 626

ODS
    ODS Graphics, 607
    Statistical Graphics Using ODS, 607
ODS destination
    ODS Graphics, 651
ODS destination FILE= option
    ODS Graphics, 645
ODS destination statement

ODS Graphics, 631, 641, 644
ODS Graphics, 607
    accessing individual graphs, 632
    axis customization, 755
    axis labels, modifying, 731
    box plot, 719, 763
    contour plot, 656
    destination, closing, 656
    destinations, 651
    diagnostics panel, 751
    DOCUMENT destination, 724
    document path, 726
    DOCUMENT procedure, 724
    Documents window, 725
    editing templates, 730
    excluding graphs, 649
    file name, base, 652
    fonts, modifying, 680
    getting started, 609
    graph label, 647
    graph modification, 634
    graph name, 647, 725
    graph resolution, 633, 656
    graph size, 633, 656
    graph template language, 687
    graph templates, 687
    graph templates, customizing, 697
    graph templates, definition, 730
    graph templates, displaying, 693
    graph templates, editing, 695
    graph templates, locating, 692
    graph templates, reverting to default, 698
    graph templates, saving, 697, 731
    graph titles, modifying, 731
    graphics image file, 651–653
    graphics image file, saving, 654
    graphics image file, type, 651
    grid lines, 737
    histogram, 722
    HTML destination, 651, 654
    index counter, 652
    LaTeX destination, 651, 655, 722
    lines, 677
    LISTING destination, 654
    markers, 677
    multiple destinations, 653, 655
    ODS destination, 651
    ODS destination FILE= option, 645
    ODS destination statement, 631, 641, 644
    ODS Graphics Editor, 658
    ODS GRAPHICS statement, 638
    PDF destination, 656
    PostScript, 655
    presentations, 720
    primer, 627
    procedures, 637
    reference lines, 755
    referring to graphs, 647
    replaying output, 725
    Results Window, 646
    RTF destination, 720
    SASHELP.Tmplmst, 697
    SASUSER.Templat, 697
    scatter plot, 689
    selecting graphs, 649, 725
    SGPANEL procedure, 711
    SGPLOT procedure, 708
    SGRENDER procedure, 714
    SGSCATTER procedure, 709
    statistical graphics procedures, 708
    style, 628, 664
    style elements, 668
    style modification, 737
    style modification, ModStyle macro, 675
    style, box plot, 762
    style, customizing, 683
    style, default, 685
    surface plot, 656
    survival plot, 612, 740
    template modification, 728
    template primary statement, 772
    template statement order, 772
    template store, default, 662
    text, adding to plots, 765
    tooltips, 719
    trace output, 728
    traditional graphics, 637
    viewing graphs, 646
ODS Graphics Editor
    ODS Graphics, 658
ODS GRAPHICS statement
    ODS Graphics, 638
ODS path, 685
ODS Statistical Graphics, *see* ODS Graphics
ODS Styles
    ANALYSIS style, 618
    DEFAULT style, 617
    JOURNAL style, 722
    JOURNAL2 style, 620
    LISTING style, 626
    RTF style, 623
    STATISTICAL style, 622
ODS template search path, 697

PDF destination
    ODS Graphics, 656
PostScript
    graphics image file, 655, 722

ODS Graphics, 655
reference lines
    ODS Graphics, 755
Results Window
    ODS Graphics, 646
RTF destination
    ODS Graphics, 720
RTF style
    ODS Styles, 623

SAS Registry, 646
SAS Registry Editor, 685
SASHELP.Tmplmst
    template store, 697
SASUSER.Templat
    template store, 697
scatter plot
    ODS Graphics, 689
SGPANEL procedure
    ODS Graphics, 711
SGPLOT procedure
    ODS Graphics, 708
SGRENDER procedure
    ODS Graphics, 714
SGSCATTER procedure
    ODS Graphics, 709
Statistical Graphics Using ODS, *see* ODS Graphics
STATISTICAL style
    ODS Styles, 622
style
    ODS Graphics, 628, 664
style elements
    ODS Graphics, 668
style modification
    ODS Graphics, 737
style modification, ModStyle macro
    ODS Graphics, 675
style, box plot
    ODS Graphics, 762
style, customizing
    ODS Graphics, 683
style, default
    ODS Graphics, 685
survival plot
    ODS Graphics, 740

Template Browser window, 730
template modification
    ODS Graphics, 728
template primary statement
    ODS Graphics, 772
template statement order
    ODS Graphics, 772

template store
    SASHELP.Tmplmst, 697
    SASUSER.Templat, 697
    user-defined, 697
template store, default
    ODS Graphics, 662
Templates window, 667, 693
text, adding to plots
    ODS Graphics, 765
tooltips
    ODS Graphics, 719
trace output
    ODS Graphics, 728

Syntax Index

ANTIALIAS= option
    ODS GRAPHICS statement, 638
ANTIALIASMAX= option
    ODS GRAPHICS statement, 639

BODY= option
    ODS HTML statement, 641
BORDER= option
    ODS GRAPHICS statement, 639

CONTENTS= option
    ODS HTML statement, 641

DOCUMENT procedure
    LIST statement, 726
    REPLAY statement, 725

FILE= option
    ODS destination statement, 645
    ODS PDF statement, 656
FRAME= option
    ODS HTML statement, 641

GPATH= option
    ODS HTML statement, 654

HEIGHT= option
    ODS GRAPHICS statement, 639

ID= option
    ODS PDF statement, 656
IMAGE_DPI= option
    ODS destination statement, 641
IMAGEFMT= option
    ODS GRAPHICS statement, 639
IMAGEMAP= option
    ODS GRAPHICS statement, 639
IMAGENAME= option
    ODS GRAPHICS statement, 639

LABELMAX= option
    ODS GRAPHICS statement, 639

MAXLEGENDAREA= option
    ODS GRAPHICS statement, 639

ODS destination statement
    FILE= option, 645
    IMAGE_DPI= option, 641
    STYLE= option, 641

ODS DOCUMENT statement, 725
ODS EXCLUDE statement, 649
ODS Graphics
    PLOTS= option, 642
ODS GRAPHICS statement
    ANTIALIAS= option, 638
    ANTIALIASMAX= option, 639
    BORDER= option, 639
    HEIGHT= option, 639
    IMAGEFMT= option, 639
    IMAGEMAP= option, 639
    IMAGENAME= option, 639
    LABELMAX= option, 639
    MAXLEGENDAREA= option, 639
    RESET= option, 640
    SCALE= option, 640
    TIPMAX= option, 640
    WIDTH= option, 641
ODS HTML statement
    BODY= option, 641
    CONTENTS= option, 641
    FRAME= option, 641
    GPATH= option, 654
    PATH= option, 654
    URL= suboption, 655
ODS LATEX statement
    STYLE= option, 722
ODS PATH statement
    RESET option, 698
    SHOW option, 697
ODS PDF statement
    FILE= option, 656
    ID= option, 656
ODS RTF statement, 721
ODS SELECT statement, 649
ODS TRACE statement, 647
    LABEL option, 648
    LISTING option, 648

PATH= option
    ODS HTML statement, 654
PLOTS= option
    ODS Graphics, 642

RESET option
    ODS PATH statement, 698
RESET= option
    ODS GRAPHICS statement, 640

SCALE= option
    ODS GRAPHICS statement, 640
SOURCE statement
    TEMPLATE procedure, 667
STYLE= option
    ODS destination statement, 641
    ODS LATEX statement, 722

TEMPLATE procedure
    SOURCE statement, 667
TIPMAX= option
    ODS GRAPHICS statement, 640

URL= suboption
    ODS HTML statement, 655

WIDTH= option
    ODS GRAPHICS statement, 641

Your Turn

We welcome your feedback.

- If you have comments about this book, please send them to `yourturn@sas.com`. Include the full title and page numbers (if applicable).
- If you have comments about the software, please send them to `suggest@sas.com`.

SAS® Publishing Delivers!

Whether you are new to the work force or an experienced professional, you need to distinguish yourself in this rapidly changing and competitive job market. SAS® Publishing provides you with a wide range of resources to help you set yourself apart. Visit us online at support.sas.com/bookstore.

SAS® Press
Need to learn the basics? Struggling with a programming problem? You'll find the expert answers that you need in example-rich books from SAS Press. Written by experienced SAS professionals from around the world, SAS Press books deliver real-world insights on a broad range of topics for all skill levels.

support.sas.com/saspress

SAS® Documentation
To successfully implement applications using SAS software, companies in every industry and on every continent all turn to the one source for accurate, timely, and reliable information: SAS documentation. We currently produce the following types of reference documentation to improve your work experience:
- Online help that is built into the software.
- Tutorials that are integrated into the product.
- Reference documentation delivered in HTML and PDF – **free** on the Web.
- Hard-copy books.

support.sas.com/publishing

SAS® Publishing News
Subscribe to SAS Publishing News to receive up-to-date information about all new SAS titles, author podcasts, and new Web site features via e-mail. Complete instructions on how to subscribe, as well as access to past issues, are available at our Web site.

support.sas.com/spn

§sas. | THE POWER TO KNOW®

SAS and all other SAS Institute Inc. product or service names are registered trademarks or trademarks of SAS Institute Inc. in the USA and other countries. ® indicates USA registration. Other brand and product names are trademarks of their respective companies. © 2009 SAS Institute Inc. All rights reserved. 518177_1US.0109